Noise Control

A Guide for Workers and Employers

Third Edition

Emory E. Knowles III, CIH, CSP
Editor

AMERICAN SOCIETY OF SAFETY ENGINEERS   DES PLAINES, ILLINOIS

Noise Control, **Third Edition** © 2003 by **The American Society of Safety Engineers.** All rights reserved. No part of this book may be reprinted or reproduced or utilized in any form or by any electronic, mechanical, or other means, including photocopying and recording, or in any information storage or retrieval system, without permission in writing from the publisher.

This publication is designed to provide accurate and authoritative information with regard to the subject matter covered. It is sold with the understanding that the publisher is not engaged in rendering legal, or other professional services. If legal or other expert assistance is required, the services of a competent professional person should be sought. While this book was carefully produced, the editor and publisher do not warrant the information contained herein to be free of errors.

Library of Congress Cataloging-in-Publication Data

Noise control : a guide for workers and employers / Emory E. Knowles, editor.-- 3rd ed.
    p. cm.
  ISBN 1-885581-44-0 (pbk. : alk. paper)
  1. Noise control--United States--Handbooks, manuals, etc. I. Knowles, Emory E. III. American Society of Safety Engineers

TD893.N65 2003
620.2'3--dc21

2003043791

Published in 2003 by
The American Society of Safety Engineers
1800 E. Oakton Street
Des Plaines, Illinois 60018-2187
   Michael F. Burditt, Manager Technical Publications

Cover and text design and composition: Michael Burditt

Printed in the United States on acid-free paper.
10 9 8 7 6 5 4 3 2 1

Foreword

"Noise Control: A Guide for Workers and Employers" was originally published by the Occupational Safety and Health Administration of the U.S. Department of Labor. The original edition was adapted and edited by Matt Witt, Director, American Labor Education Center, from a publication of the Swedish Work Environment Fund, and was printed with the Fund's permission.

A second edition of the book was edited in 1984 by Richard Stepkin, MS, C.C.C.-A, an industrial audiologist and President of Enviromed Corporation in Laurel Springs, New Jersey and Ralph E. Mosely, MBS, CSP, President, Mosely and Associates, Ltd., Nashville, Tennessee. The second edition was published by ASSE. Stefan Weisz, CSP of the Potomac Electric Power Company, Washington, DC also contributed to the second edition.

The American Society of Safety Engineers wishes to extend its sincere appreciation to the Editor of this new edition, Emory Knowles, and to Darrick L. Bertig, CIH, CSP, CUSA, and George Byrns, Ph.D., MPH, CIH, for their valuable comments and suggestions in reviewing the draft of the third edition.

The ASSE recognizes the value of this publication and by updating, expanding, and re-issuing this book, it wishes to continue it's availability and usefulness. To this edition is added an explanation of the mechanism of hearing in "How We Hear", a revision of the appendix detailing OSHA requirements relative to noise, as well as complete listings of federal and state OSHA offices in the appendices.

"Noise Control: A Guide for Workers and Employers" is not intended as a legal interpretation of the Occupational Safety and Health Act of 1970 or of OSHA or any state-plan standards.

About the Editor

Emory E. Knowles has over 23 years of experience in safety and health management and is currently Manager, Industrial Hygiene and Safety for the Northrop Grumman Corporation Electronic Systems sector, headquartered in Baltimore, Maryland. His responsibilities include management of safety, industrial hygiene, and product safety programs at more than 50 domestic and international sites.

For many years he has been an adjunct faculty member teaching industrial hygiene and occupational safety courses at Catonsville Community College in Baltimore, Maryland and is a faculty Associate at Johns Hopkins University, where he teaches a course on Occupational Safety in the Graduate School of Hygiene and Public Health.

Mr. Knowles is both a Certified Safety Professional and a Certified Industrial Hygienist and a Professional Member of ASSE. He has a Bachelor of Arts degree in Biology from the College of Steubenville, a Certificate in Occupational Safety and Health from Catonsville Community College, a Master of Science degree in Microbiology from West Virginia University, and a Master of Science degree in Safety Sciences from Indiana University of Pennsylvania.

Mr. Knowles has numerous publications in the fields of microbiology, safety, and industrial hygiene. In 1997 he received the national Johnson and Higgens Scrivener Editorial Award for the most noteworthy article in the field of safety that year. In 2000 he was presented the Charles Culbertson Outstanding Volunteer Service award. He was a past "Safety Professional of the Year" at both the Chapter and Region levels.

He is active on the local, regional, and national level in the ASSE and is currently Chair, ASSE Governmental Affairs Committee and is a member of the Council on Professional Affairs. He also serves on the Board of Directors of the Safety Council of Maryland, the Safety Council of Maryland Executive Committee, and is the Vice-President for southern Maryland.

He and his wife Jennie have three children and reside in Baltimore County, Maryland.

Table of Contents

Foreword
About the Editor

Introduction   1

Noise Control: Basic concepts and terminology   5

Application of Noise Control Principles   11

    Sound Behaviour, 12
    Sound from Vibrating Plates, 29
    Sound Production in Air or Gases, 45
    Sound Production in Flowing Liquids, 59
    Sound Movement Indoors, 63
    Sound Movement in Ducts, 73
    Sound from Vibrating Machines, 83
    Sound Reduction in Enclosure Walls, 95

Noise Control Measures   103

    Changes in Machinery and Equipment, 103
    Materials Handling, 105
    Enclosure of Machines, 106
    Control of Noise from Vibrating Surfaces, 107
    Damping with Absorbents, 108
    Sound Insulating Separate Rooms, 109
    Maintenance, 110
    Planning for Noise Control, 110

What is Required in the OSHA Standards?   113

    Legal Limits on Noise, 113
    Monitoring, 113
    Audiometric Testing Program, 115

Hearing Protectors, 115
Training Program, 116
Consultation for Employers, 116
Loans for Small Businesses, 116
Workers' Rights Under OSHA, 116
OSHA Occupational Hearing Loss Recordkeeping, 118
How to Make Noise Measurements, 119
How to Solve Noise Problems, 120

Appendix A—Sources of Information and Further Reading   **123**

Appendix B—Government Agencies   **125**

Appendix C—OSHA Noise Standard Title 29 Code of Federal Regulations (CFR) 1910.95 - Occupational Noise Exposure   **129**

1910.95 Appendix A - Noise Exposure Computation, 142
1910.95 Appendix B - Methods for Estimating the Adequacy of Hearing Protector Attenuation, 149
1910.95 Appendix C - Audiometric Measuring Instruments, 151
1910.95 Appendix D - Audiometric Test Rooms, 152
1910.95 Appendix E - Acoustic Calibration of Audiometers, 153
1910.95 Appendix F - Calculations and Application of Age Corrections to Audiograms, 155
1910.95 Appendix G - Monitoring Noise Levels—Non-mandatory Information, 160
1910.95 Appendix H - Availability of Referenced Documents, 163
1910.95 Appendix I - Definitions, 164

Appendix D—Part 1904 - OSHA Recordkeeping Related to Hearing Loss Cases   **167**

Appendix E—Directory of OSHA Regional Offices   **171**

Appendix F—Directory of States with Approved Occupational Safety and Health Plans   **175**

Appendix G—Directory of OSHA Consultation Offices   **181**

Introduction

Noise—what can it do? It can destroy hearing. It can create physical and psychological stress. It can contribute to incidents by making it difficult or impossible to hear warning signals.

Millions of workers wordwide are exposed to hazardous noise. But, it is important to note that *noise exposure can be controlled.* No matter what the noise problems may be in a particular workplace, technology can be applied to reduce the hazard.

It may be possible to:

- Use quieter work processes.
- Alter or enclose equipment to reduce noise at *the source.*
- Use sound-absorbing materials to prevent the *spread of noise by isolating the source.*

This book is presented by for workers and employers interested in reducing workplace noise. The American Society of Safety Engineers (ASSE) believes that highly technical training is not generally necessary in order to understand the basic principles of noise control. Noise problems can often be solved by the workers and employers who are directly affected.

The book contains four major sections:

First, a brief overview of the effects of noise on human health and a discussion of some of the key words and concepts involved in noise control.

Second, an explanation of specific principles of noise control which the reader can apply in his or her workplace.

Third, a discussion of particular techniques for controlling noise.

Fourth, a description of the ways OSHA can help employers and employees, including an explanation of the legal requirements for noise control which employers must follow.

ASSE hopes that the information in this book will be discussed by employers, workers, and union representatives.

How Humans Hear Sound

Sound is propagated through the atmosphere as a wave form of energy. When sound reaches the ears, the waves are funneled via the outer ear to the eardrum, a thin membrane that separates the outer from the middle ear (see the drawing of the human ear on the next page). The sound waves cause the eardrum to vibrate and the vibrations are transferred to three small bones (the ossicles) in the middle ear commonly termed the hammer, the anvil, and the stirrup.

As these bones vibrate, the vibrations are transmitted to the fluid-filled inner ear causing movement in the fluid. Within the fluid and lining of the inner ear are small hair cells with hair-like nerve endings extending into the fluid. As the vibrations are transmitted through the fluid, the hair-like structures of the hair cells move. These movements are transformed into nerve impulses in the auditory nerve.

Hearing loss generally can occur due to damage to the structures of the ear or damage to the hair cells. Damage to the ear structures, such as the tiny bones of the middle ear or a punctured ear drum is termed *conductive hearing loss* because the hearing loss is caused by damage to structures that capture the sound vibrations. This type of loss could be the result of a blow to the head or an explosion. Damage to the hair-like structures of the hair cells results in actual damage to the nerve endings of the auditory nerve and this type of loss is termed *sensorineural hearing loss*. Sensorineural hearing loss results when the hair-like structures are repeatedly assaulted by high noise levels causing permanent loss in the ability to transmit the sound as a nerve impulse.

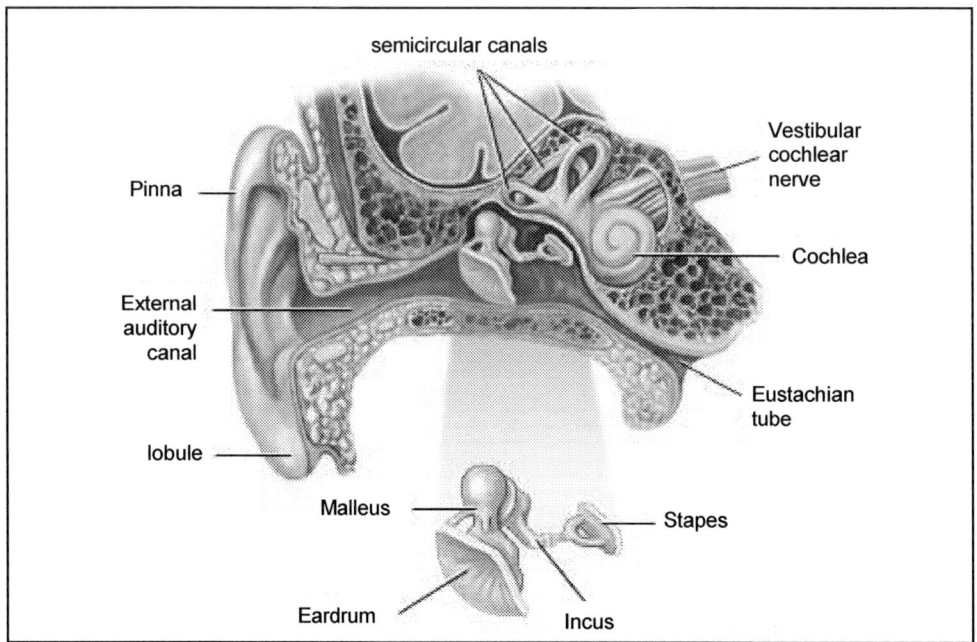

Figure 1. Anatomy of the human ear.

Generally, conductive hearing loss can be medically corrected. For most sensorineural hearing losses, the loss is permanent. We all lose some hearing as a result of the aging process. Loss of hearing due to aging is termed *presbycusis*.

Noise: Its effect on health

The ability to hear is one of our most precious gifts. Without it, it is very difficult to lead a full life either on or off the job. Excessive noise can destroy the ability to hear, and may also put stress on other parts of the body, including the heart. *For most effects of noise, there is no cure,* so that prevention of excessive noise exposure is the only way to avoid health damage.

Hearing

The damage done by noise depends mainly on how loud it is and on the length of exposure. The frequency or pitch can also have some effect, since high-pitched sounds are more damaging than low-pitched ones.

Noise may tire out the inner ear, causing temporary hearing loss. After a period of non-exposure, hearing may be restored. Some workers who suffer temporary hearing loss may find that by the time their hearing returns to normal, it is time for another work shift, so, in that sense, the problem is "permanent."

With continual noise exposure, the ear will lose its ability to recover from temporary hearing loss, and the damage will become permanent. Permanent hearing loss results from the destruction of cells in the inner ear-cells which can never be replaced or repaired. Such damage can be caused by long-term exposure to loud noise or, in some cases, by brief exposures to **very** loud noises.

Normally, workplace noise first affects the ability to hear high frequency (high-pitched) sounds. This means that even though a person can still hear some noise, speech or other sounds may be unclear or distorted. Workers with hearing impairment typically say, "I can hear you, but I can't understand you." Distortion occurs especially when there are background noises or many people talking. As conversation becomes more difficult to understand, the person becomes isolated from family and friends. Music and the sounds of nature become impossible to enjoy. A hearing aid can make speech louder, but cannot make it clearer, and is rarely a satisfactory remedy for hearing loss. Workers suffering from noise-induced hearing loss may also experience continual ringing in their ears, called "tinnitus." At this time, there is no cure for tinnitus, although some doctors are experimenting with treatment.

Other Effects

Although research on the effects of noise is not complete, it appears that noise can cause quickened pulse rate, increased blood pressure and a narrowing of the

Noise Control

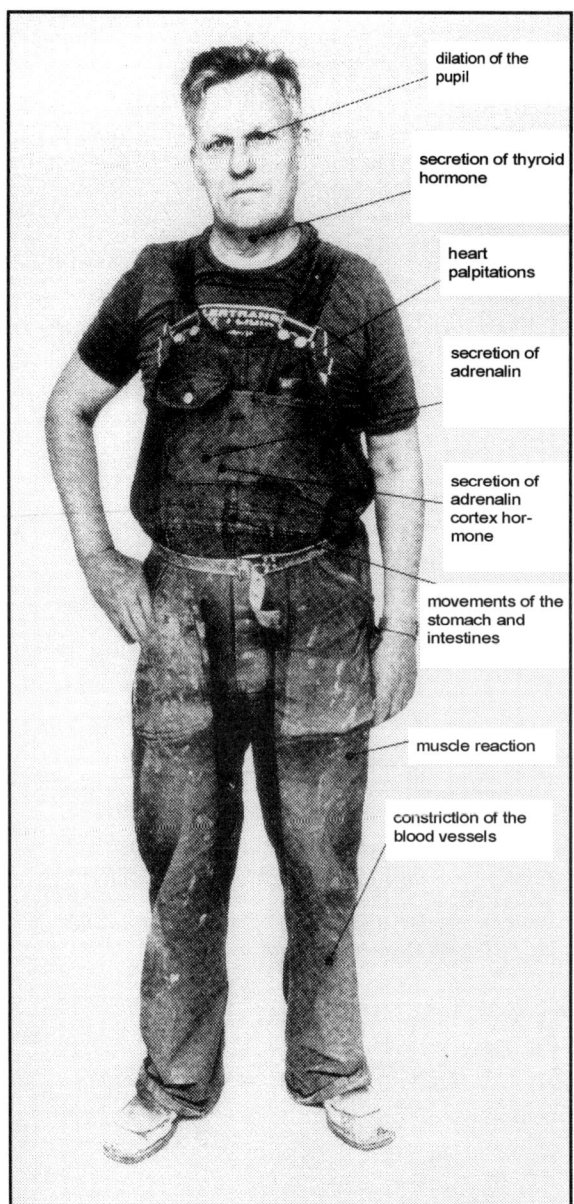

Figure 2. In addition to causing hearing loss by destroying the inner ear, noise apparently can put stress on other parts of the body by causing reactions such as those shown.

blood vessels. Over a long period of time, these may place an added burden on the heart.

Noise may also put stress on other parts of the body by causing the abnormal secretion of hormones and tensing of muscles (see Figure 2). Workers exposed to noise sometimes complain of nervousness, sleeplessness and fatigue. Excessive noise exposure also can reduce job performance and may cause high rates of absenteeism.

Noise Control:
Basic concepts and terminology

There are a number of words and concepts which must be understood before beginning a discussion of noise control methods.

Sound

Sound is produced when a sound source sets the air nearest to it in wave motion. The motion spreads to air particles far from the sound source. Sound travels in air at a speed of about 340 meters per second. The rate of travel is greater in liquids and solids; for example, 1,500 m/s in water and 5,000 m/s in steel.[1]

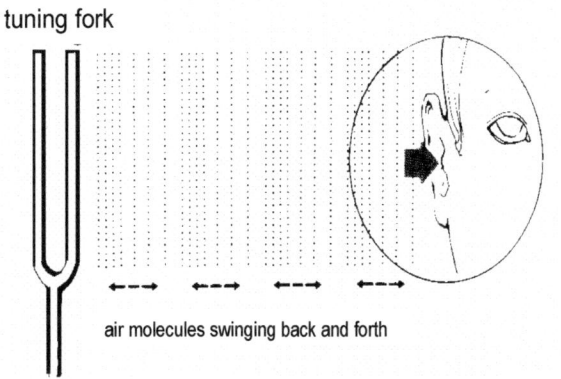

Figure 2. The sound source vibrates and affects air molecules, which strike the ear drum.

Frequency (Hz)

The frequency of a sound wave refers to the number of vibrations per second, measured in units of hertz (Hz). Sound is found within a large frequency range; audible sound for young persons is between about 20 Hz and 20,000 Hz.

The boundary between high and low frequencies is generally established at 1,000 Hz. Most human speech falls between 250 and 8,000 Hz.

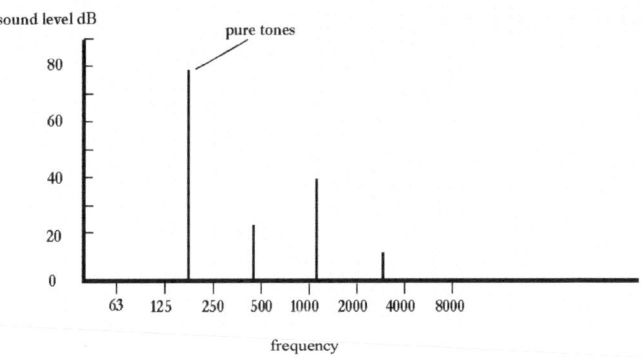

Figure 3. A pure tone is marked by a single column indicating the frequency and the sound level or intensity. Musical notes contain several tones of different frequencies and intensities.

Sound may consist of a single pure tone, but in general it is made up of several tones of varying intensities.

Noise

It is customary to call any undesirable sound "noise." The disturbing effects of noise depend both on the intensity and the frequency of the tones. For example, higher frequencies are more disturbing than low ones. Pure tones are more disturbing than a sound made up of many tones.

Infrasound and Ultrasound

Sound with frequencies below 20 Hz is called infrasound, and sound with more than 20,000 Hz is called ultrasound. There is some evidence that these sounds which cannot be heard can under certain conditions be hazardous to workers' health. This book deals only with noise that can be heard and that may damage hearing. (Refer to the latest edition of the American Conference of Governmental Industrial Hygienists' Threshold Limit Values for more information on infrasound and ultrasound.)

Decibel (dB)

Sound levels are measured in units of decibels (dB). If sound is intensified by 6 dB, it seems to the ears approximately as if the sound intensity has doubled. A reduction by 6 dB makes it seem as if the intensity has been reduced by half.

Noise Level Measurement

In measuring sound levels, instruments are used which resemble the human ear in

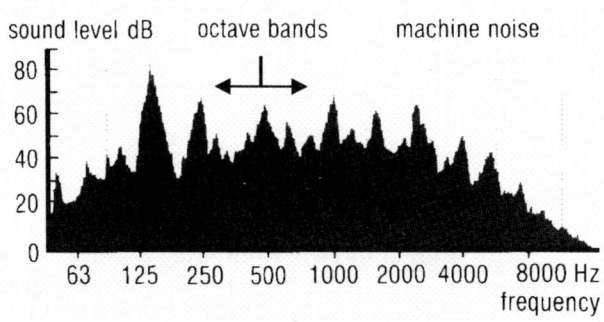

Figure 5. Noise is a disorderly mixture of tones at many frequencies.

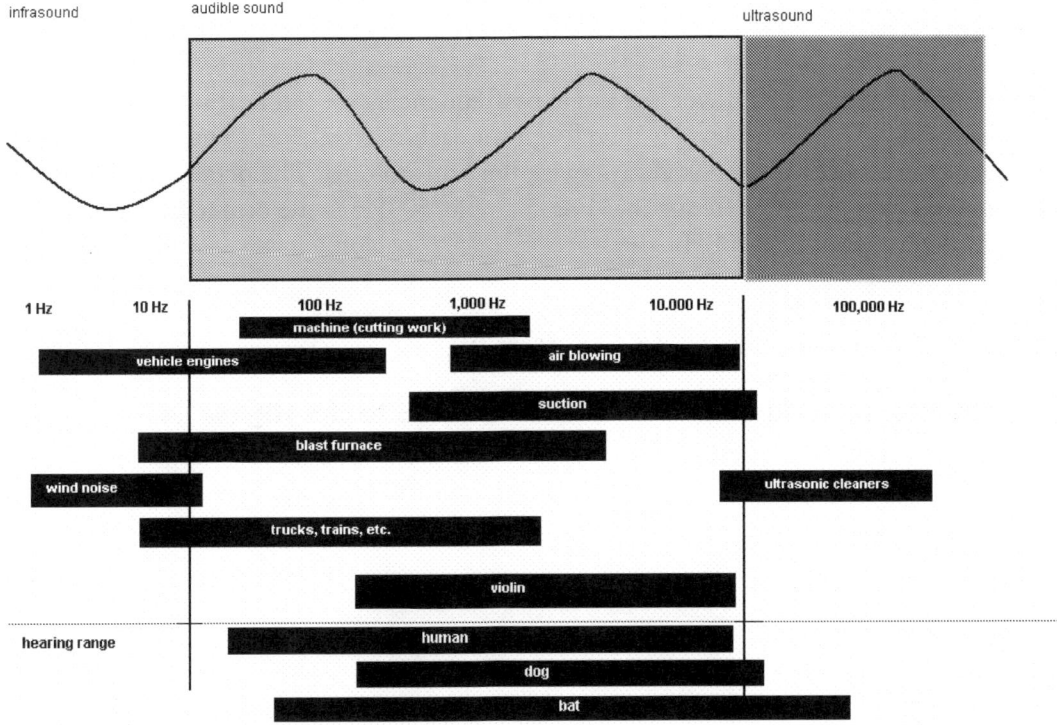

Figure 6. At the same intensity, the noise from a truck is less disturbing than the sound of air blowing or suction because it is at a lower frequency.

sensitivity to noise composed Of varying frequencies. The instruments measure the "A-weighted sound level" in units called dB(A).

Workplace noise measurements indicate the combined sound levels of noise from a number of sources (machinery and materials handling) and background noise (from ventilation systems, cooling compressors, circulation pumps, etc.).

In order to accurately identify all workplace noise problems, the noise from each source should also be measured separately. Measurements at various production rates may be useful in considering possible control measures. A number of manuals for noise measurements are commercially available.

Adding Noise Levels

Decibel levels for two or more sounds cannot simply be added. Table 1 shows how the combined effect of two sounds depends on the difference in their levels. Two or more sounds of the same level combine to make a higher noise level.

Octave Band

It is common practice to divide the range of frequencies we can hear into eight octave bands. The sound level is then listed for each octave band. The top frequency in an octave band is always twice the bottom one. The octave band may be referred to by a center frequency. For example, 500Hz is the center frequency for the octave band 354-708Hz.

Table 1. Adding Noise Levels of Two or More Sounds

Differences in dB values	**Add to the Higher Level**
0-1	3 dB
2-3	2 dB
4-9	1 dB
10 or more	0 dB

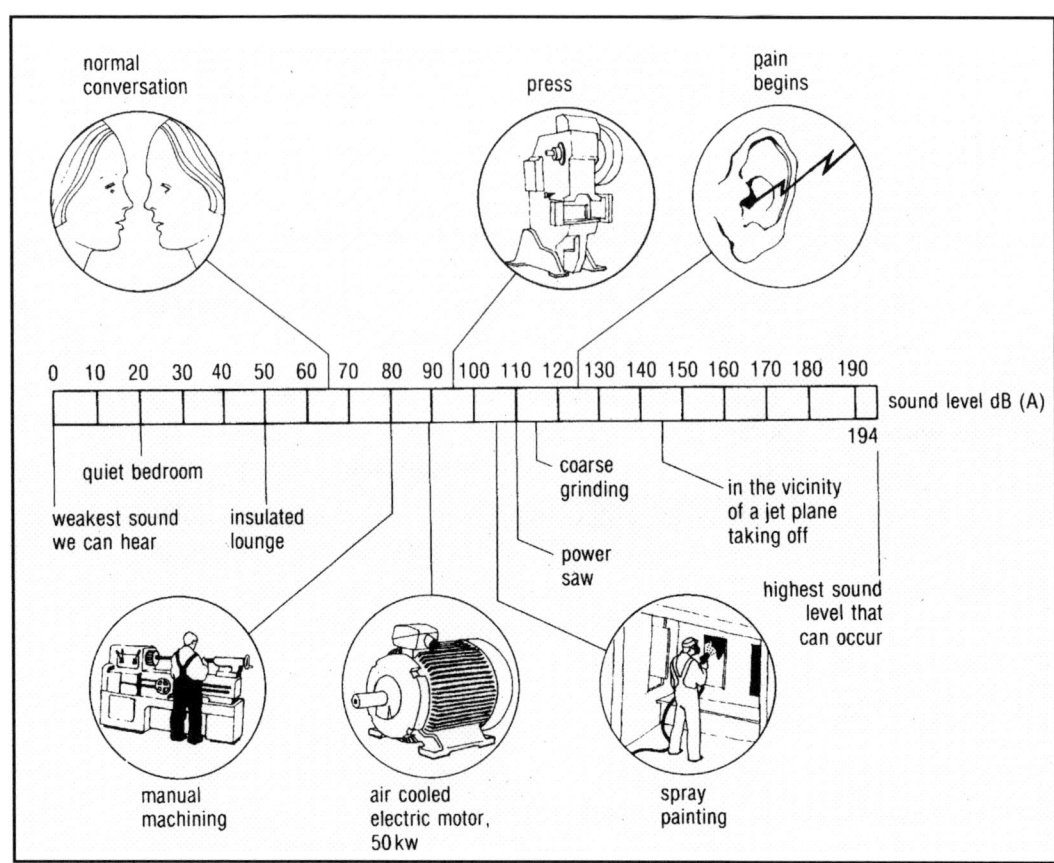

Figure 7. Comparison of noise levels, dB(A).

Sound Transmission

The word "sound" usually means sound waves traveling in air. However, sound waves also travel in solids and liquids. These sound waves may be transmitted to air to make sound we can hear.

Resonance

Each object or volume of air will "resonate," or strengthen a sound, at one or more particular frequencies. The frequency depends on the size and construction of the object or air volume.

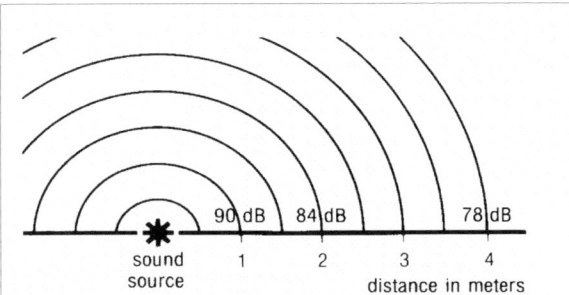

Figure 8. If a small sound source produces a sound of 90 dB at a distance of 1 meter, the sound level at a 2 meter distance is 84 dB, at 4 meters 78 dB, etc., assuming no reverberation.

Sound Reduction by Distance

Sound spreading in open air and measured at a certain distance from the source is reduced by about 6 dB for each doubling of that distance. Sound is reduced less when spreading inside a room due to reverberation (Figure 8).

Sound Transmission Loss (TL)

When a wall is struck by sound, only a small portion of the sound is transmitted through the wall, while most of it is reflected (Figure 9). The wall's ability to block transmission is indicated by its transmission loss (TL) rating, measured in decibels. The TL of a wall does not vary regardless of how it is used.

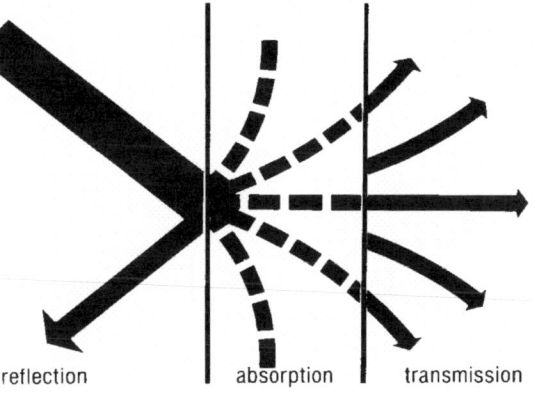

Figure 9. Part of the sound that strikes a wall is reflected, part is absorbed, and part is transmitted. The transmission loss (TL) of the wall is determined by the portion of the noise which is not transmitted through the wall.

Noise Reduction (NR)

Noise reduction is the number of decibels of sound reduction actually achieved by a particular enclosure or barrier. This can be measured by comparing the noise level before and after installing an enclosure over a noise source. NR and TL are not necessarily the same.

Sound Absorption

Sound is absorbed when it strikes a porous material. Commercial sound-absorbing materials usually absorb 70 percent or more of the sound that strikes them.

Application of Noise Control Principles

The following section explains how to apply basic noise control principles. In many cases, several principles must be applied and several control measures must be taken. Of course, these principles do not cover every possible noise problem.

The principles are discussed in eight sections:
1. Sound behavior
2. Sound from vibrating plates
3. Sound production in air or gases
4. Sound production in flowing liquids
5. Sound movement indoors
6. Sound movement in ducts
7. Sound from vibrating machines
8. Sound reduction in enclosure walls.

Various symbols are used throughout the drawings. Large black arrows indicate strong sound radiation and smaller arrows indicate reduced sound radiation.

Application of Noise Control Principles

Sound Behavior

Changes in Force, Pressure, or Speed Produce Noise

Sound is always produced by changes in force, pressure, or speed. Great changes produce louder noises and small changes quieter ones. More noise is produced if a task is carried out with greater force for a short duration than with less force for a longer duration.

Principle

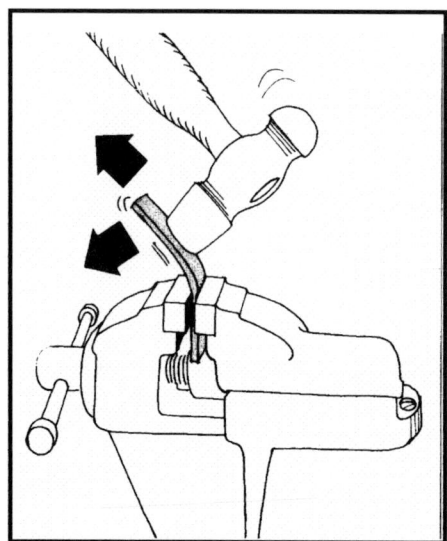 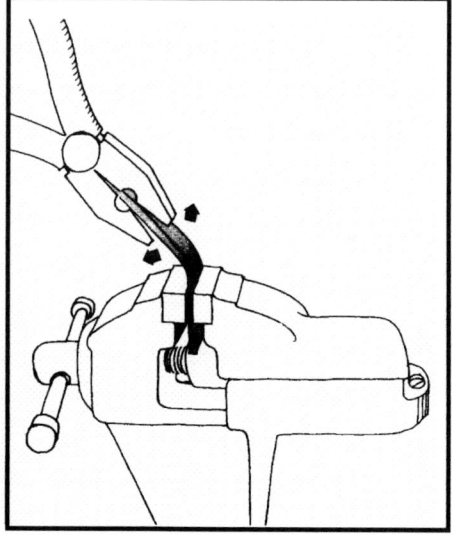

A flat strip of metal can be bent noisily by striking with a hammer,

...or quietly by bending with pliers.

Example

In a box machine, cardboard is cut with a knife blade. The knife must cut very rapidly and with great force in order for the cut to be perpendicular to the strip. High noise levels result.

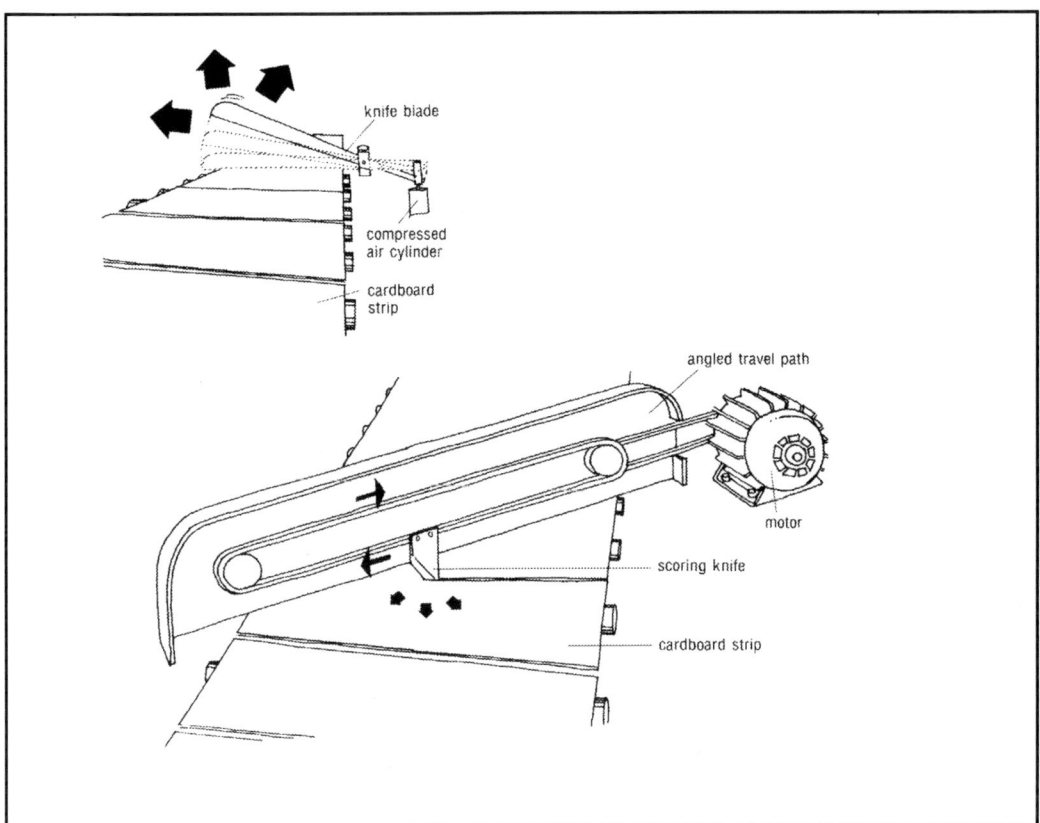

Control Measure

Using a blade that travels across the strip, the cardboard can be scored with minimal force for a longer time. Since the cardboard strip continues to move, the knife must travel at an angle in order for the cut to be perpendicular. The cutting is practically noise free.

Airborne Sound is Usually Caused by Vibration in Solids or Turbulence in Fluids

As an example of this principle, vibrations of the strings in a bass musical instrument are transmitted through the bridge to the sound box. When the sound box vibrates, sound is transmitted to the air. Another example is exemplified by a circulation pump that produces pressure variations in the water in a heating system. The sound waves are transmitted through the pipes to the radiators whose large metal surfaces transmit airborne sound.

Principle

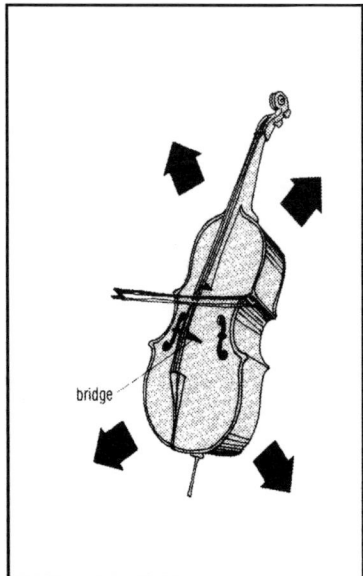
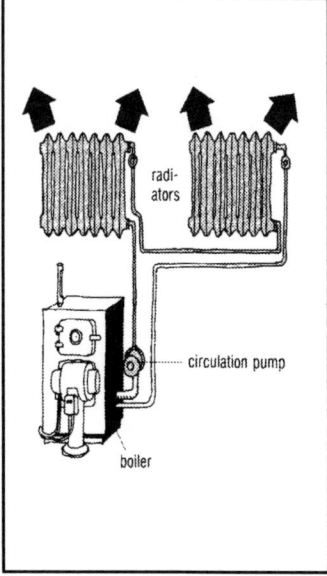

Example

Turbulent fluid flow within pipes produces sound which can be radiated through the pipes and even transmitted to the building structure by way of mounting brackets.

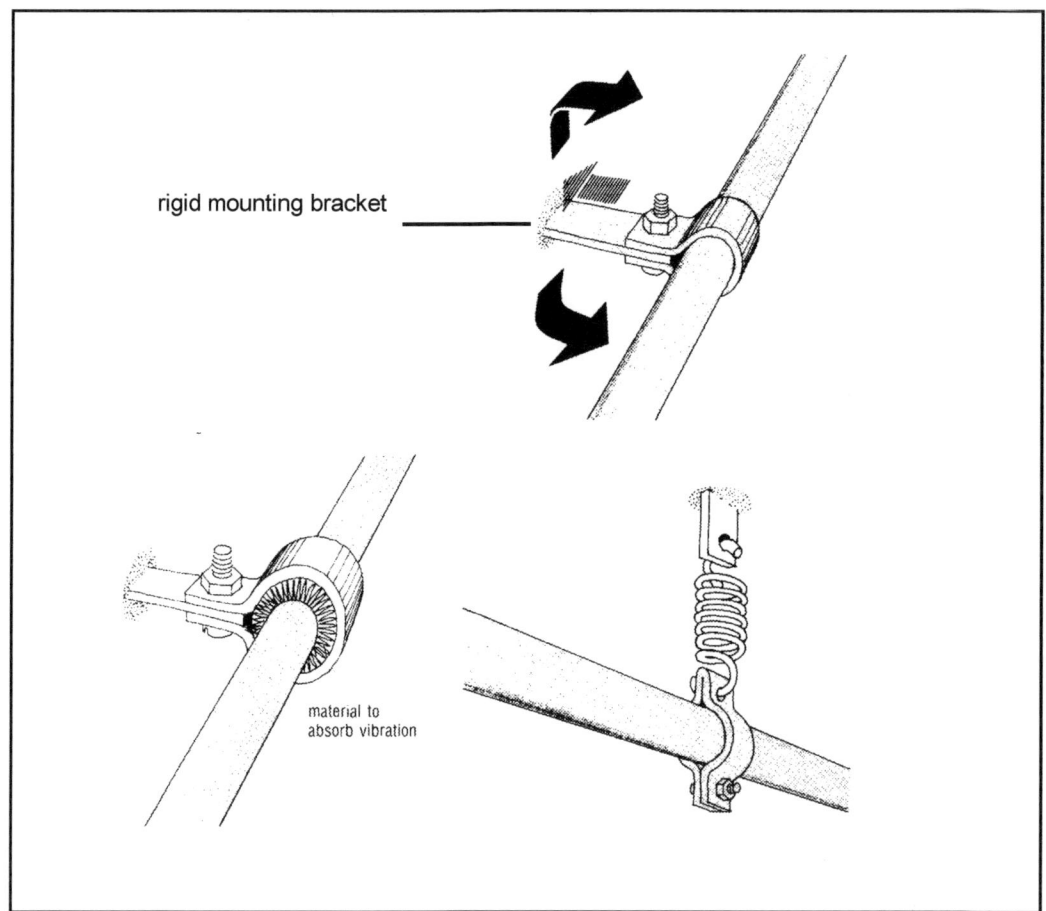

Control Measure

In addition to reducing the turbulence in the pipe by, for example, reducing flow or eliminating sharp bends, the pipe can be covered with sound absorbing material. The vibrations can also be isolated from the wall or ceiling with flexible connecting mechanisms, such as springs.

Application of Noise Control Principles

Vibrations Can Produce Sound After Traveling Great Distances

Vibrations in solids and liquids can travel great distances before producing airborne sound. Such vibrations can cause distant structures to resonate. The best solution is to stop the vibration as close to the source as possible.

Example

Vibrations from a train move along the rails and can be heard a considerable distance away.

Example

Vibrations from an elevator are transmitted throughout a building.

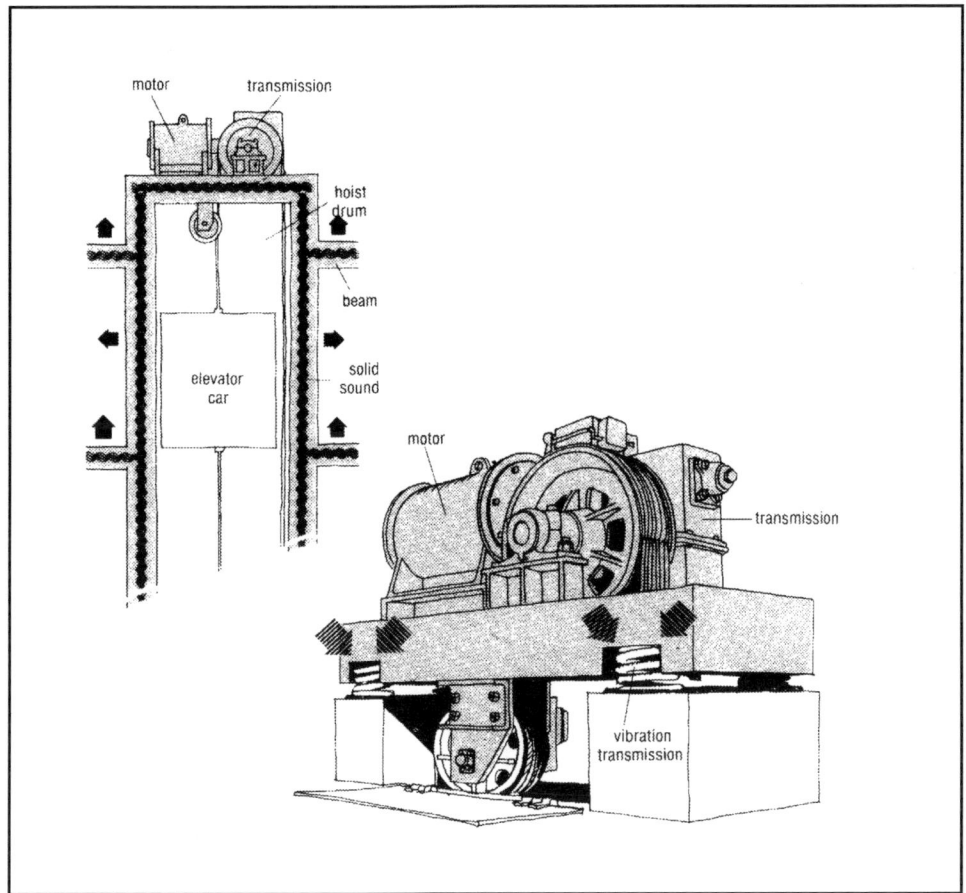

Control Measure

The elevator can be isolated from the building structure with heavy-duty springs.

The Slower the Repetition, the Lower the Frequency of the Noise

The level of low frequency noise from a sound source is determined primarily by the rate at which the changes in force, pressure, and speed are repeated. The longer the time between changes, the lower the frequency of the noise generated. The level of noise depends on the amount of the change.

Principle

The exhaust from a slowly revolving tugboat engine produces a soft, thudding, low-frequency noise.

An outboard motor's rapidly repeated pressure shocks produce a higher-frequency sound.

Example

Two gears have the same pitch diameter but different numbers of teeth. If they rotate at the same speed, the gear with fewer teeth will produce a lower-frequency noise.

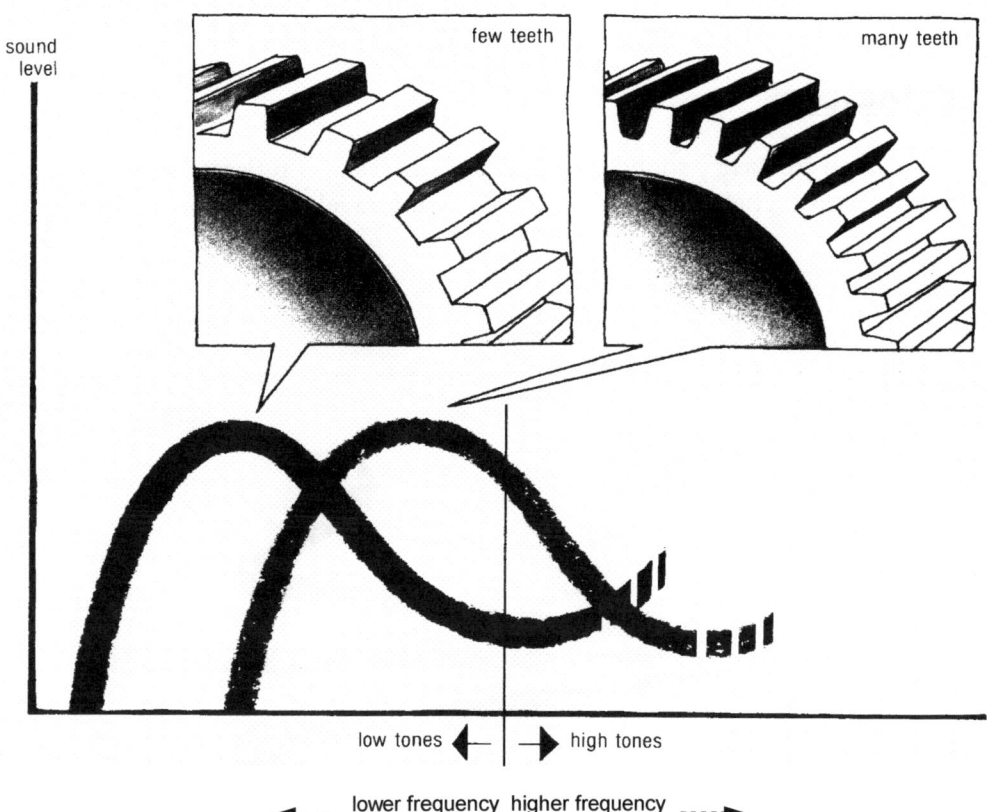

High Frequency Sound is Strongly Directional and More Easily Reflected

Principle

When high-frequency sound strikes a hard surface, it is reflected much like light from a mirror. High-frequency sound does not travel around corners easily.

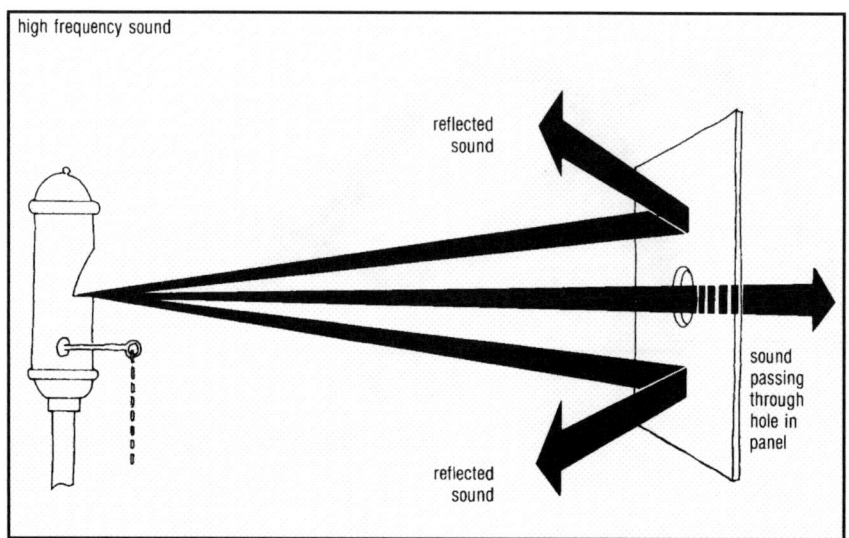

Example

High frequency noise travel directly from a high-speed riveting machine to a worker's ears.

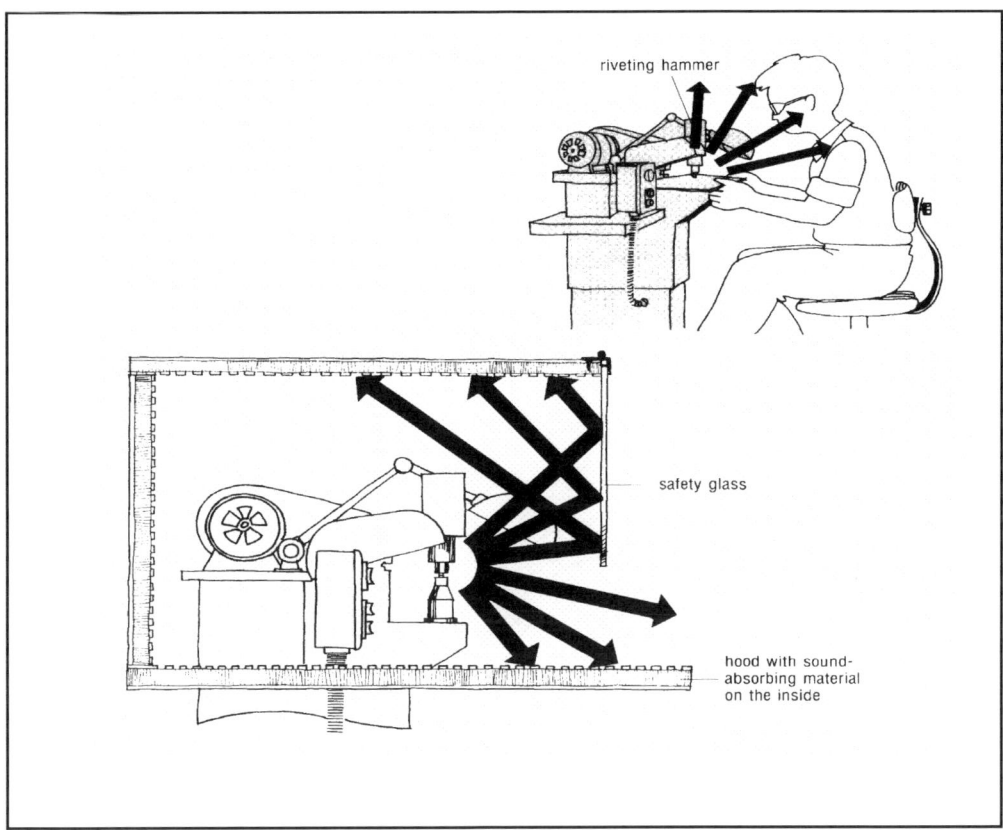

Control Measure

A sound-insulating hood, open toward the bottom of the machine, is constructed above the hammer. The hood is coated on the inside with sound-absorbing material. The upper portion of the opening is covered with safety glass. As sound starts towards the ears, the glass reflects it against the sound-absorbing walls. The sound level for the machine operator is thus reduced.

Low Frequency Noise Travels Around Objects and Through Openings

Low frequency noise radiates at approximately the same level in all directions. It travels around corners and through holes, and then continues to travel in all directions. A shield has little effect unless it is very large.

Principle

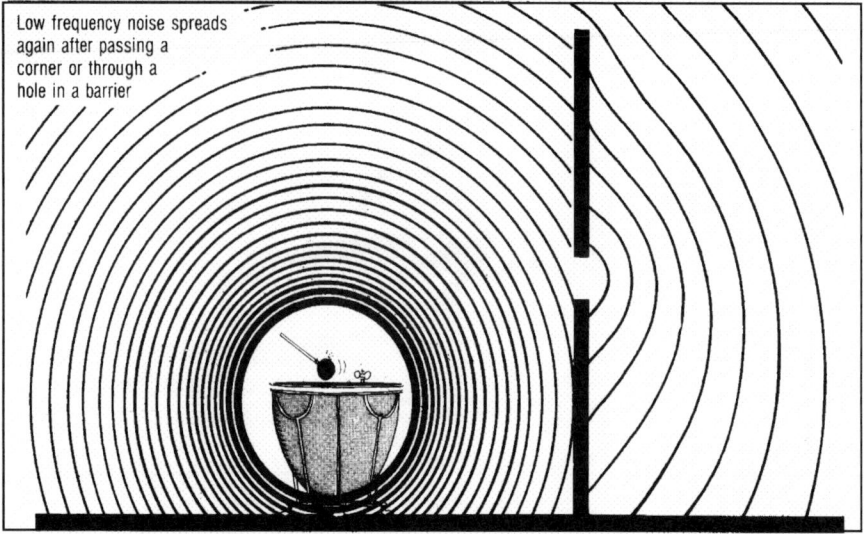

Example

Compressors and their internal diesel engines may produce strong low frequency noise, even when provided with effective mufflers at the intake and exhaust.

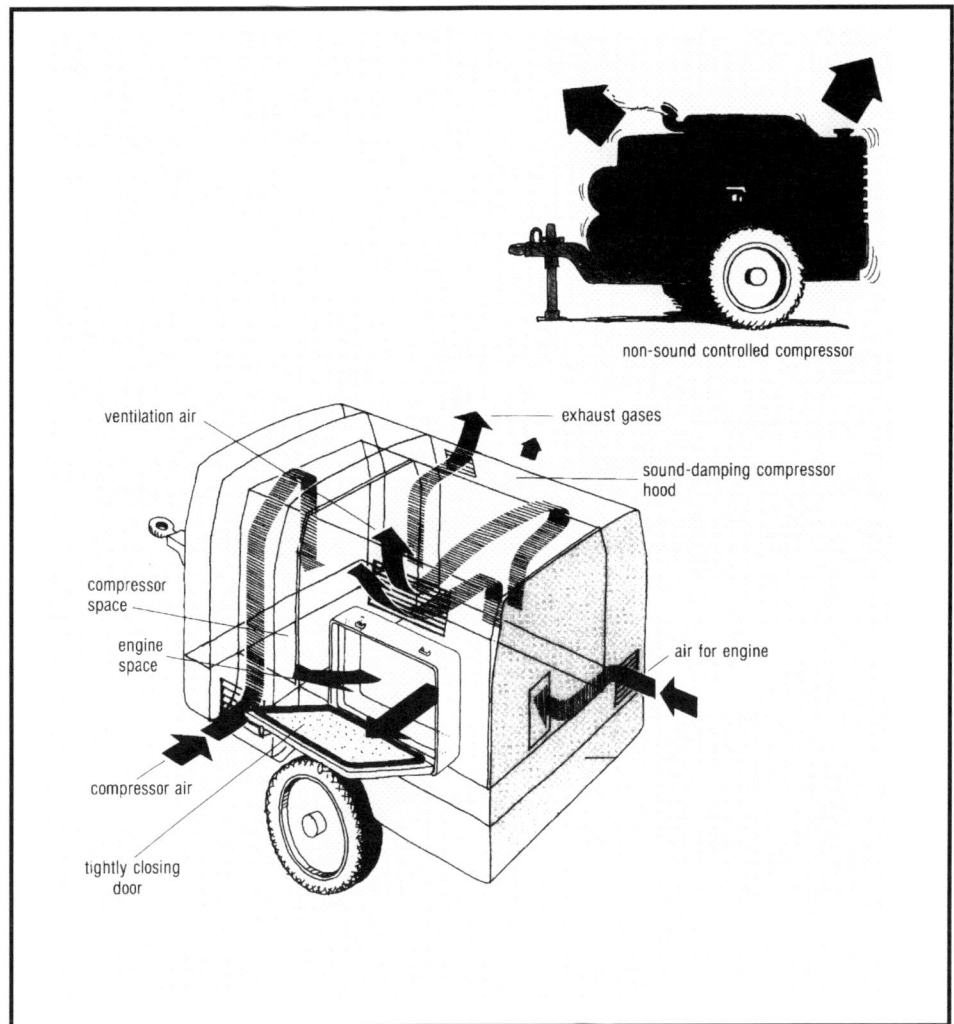

Control Measure

A complete enclosure of damped material lined with sound absorbent will reduce noise. The air and exhaust gases must pass through mufflers which are partly made of channels with sound absorbing walls. Doors for inspection must close tightly.

High Frequency Sound is Greatly Reduced by Passing Through Air

High-frequency sound is reduced or attenuated more effectively than low-frequency sound by passing through air. In addition, it is easier to insulate and shield high-frequency noise. If the noise source does not cause problems in its immediate vicinity, it may be worthwhile to shift the sound toward higher frequencies so that it can be insulated or shielded more effectively.

Principle

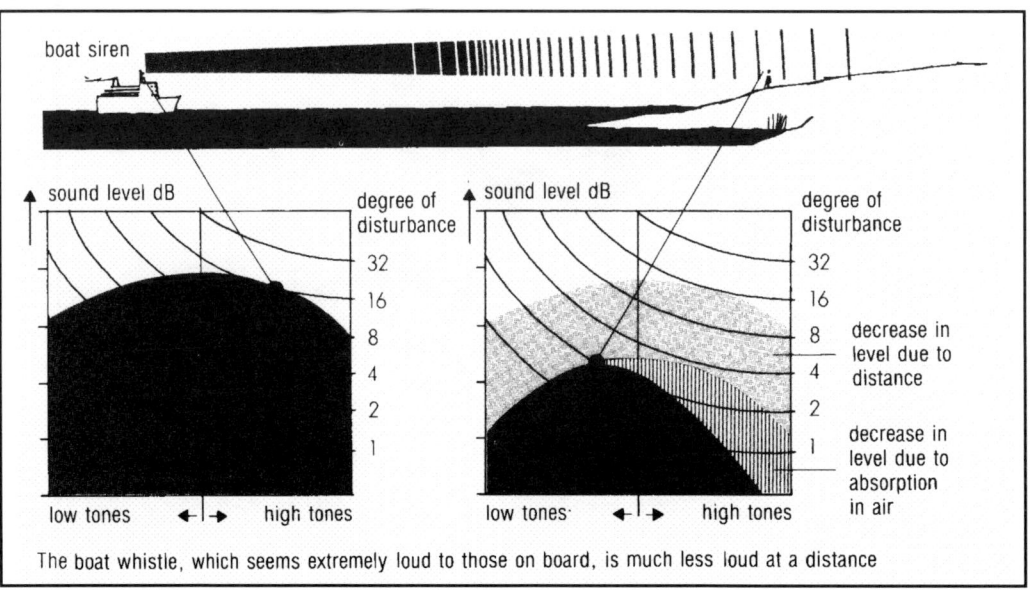

The boat whistle, which seems extremely loud to those on board, is much less loud at a distance

Example

The low-frequency noise from roof fans in an industrial building disturbs residents of houses a quarter-mile away.

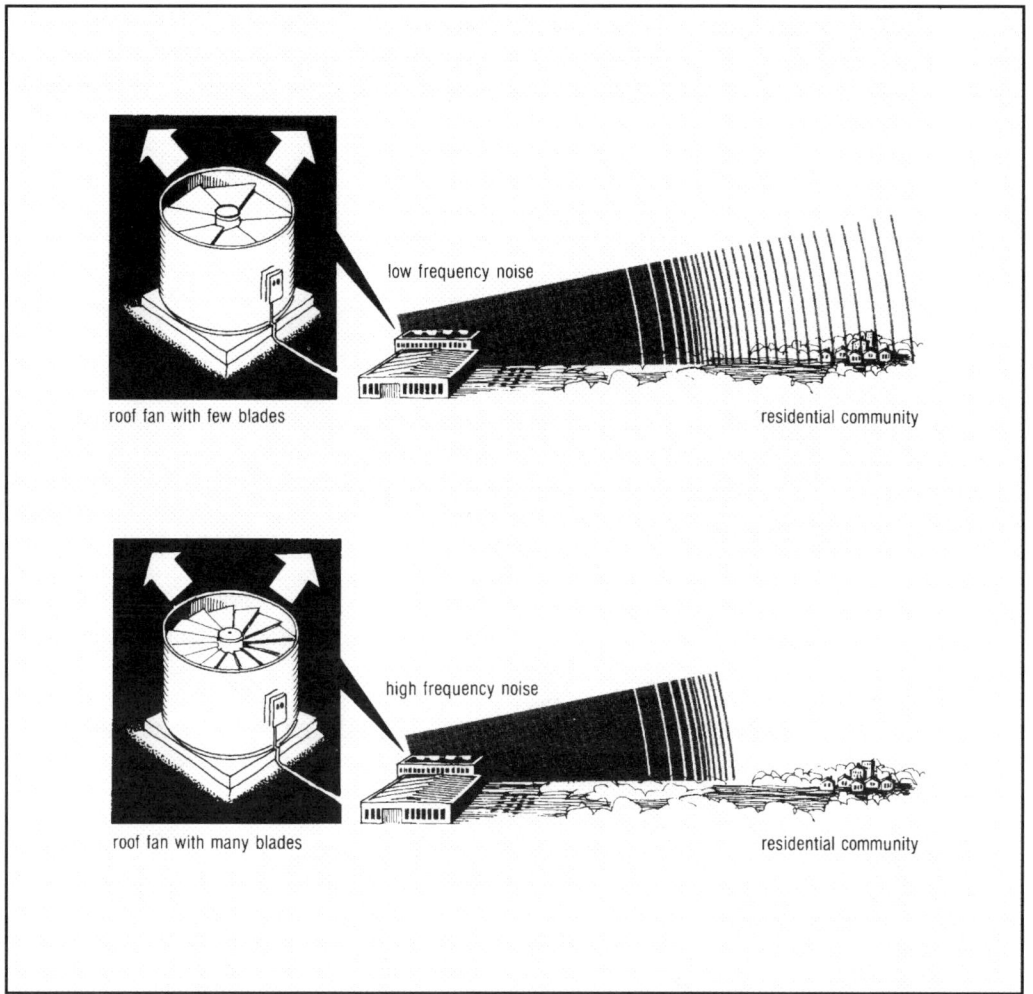

Control Measure

The rooftop fan is replaced by another one of similar capacity but with a larger number of fan blades. This produces less low-frequency noise and more high-frequency noise. The low –frequency noise no longer causes disturbances, and the high-frequency noise is adequately reduced by distance.

Low Frequency Noise is Less Disturbing

The human ear is less sensitive to low-frequency noise than to high-frequency noise. If it is not possible to reduce the noise, it may be possible to change the noise source so that more of the noise is at lower frequencies. It is important to note the high levels of low-frequency noise are more difficult to control than high-frequency noise.

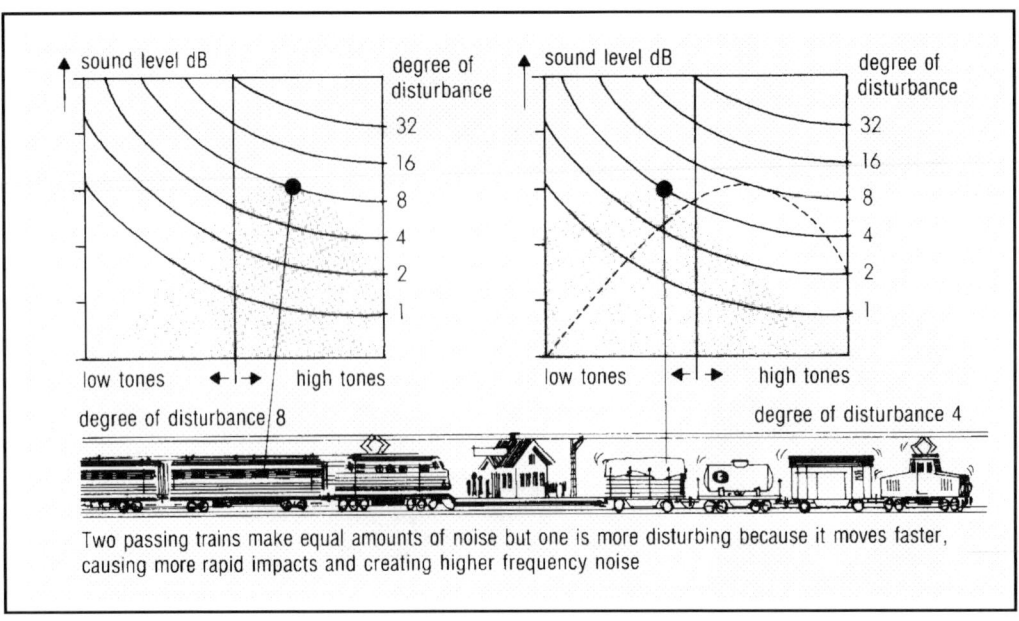

Two passing trains make equal amounts of noise but one is more disturbing because it moves faster, causing more rapid impacts and creating higher frequency noise

Example

The diesel engine in a ship operates at 125 rpm and is directly connected to the propeller. The noise from the propeller is extremely disturbing on board.

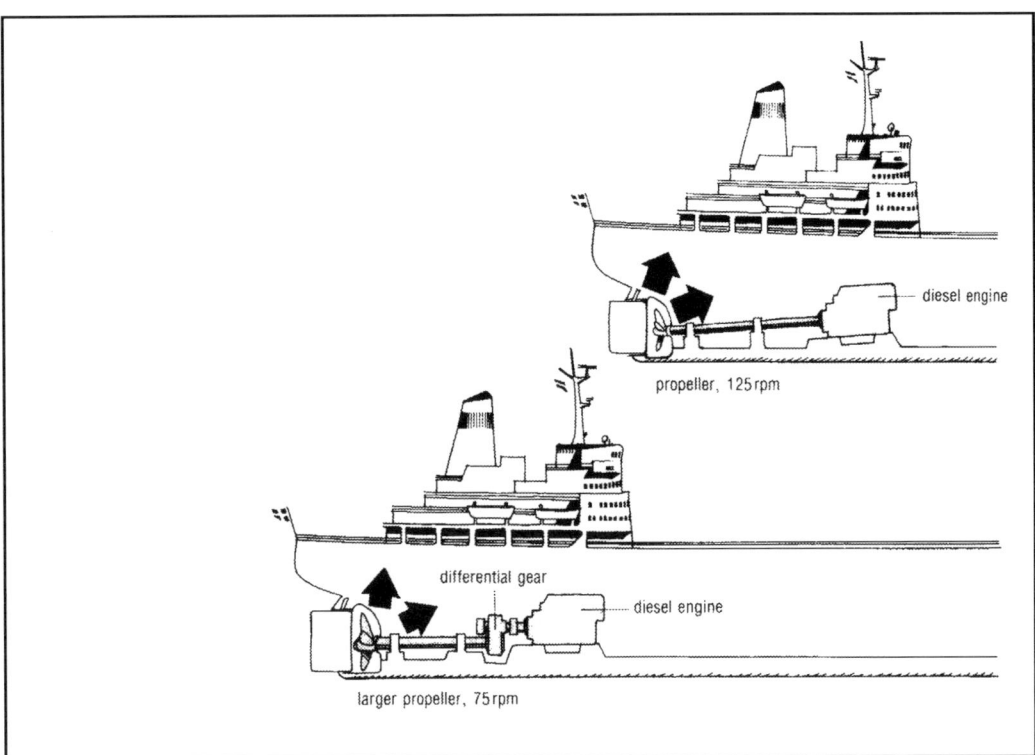

Control Measure

A reduction gear is installed between the motor and the propeller so that the propeller will revolve at 75 rpm, thereby shifting the noise to a lower frequency and making it less disturbing. The propeller is replaced by a larger one to maintain a sufficient service speed.

Sound from Vibrating Plates

Small Vibrating Surfaces Give Off Less Noise Than Large Ones

An object with a smaller surface area may vibrate intensely without a great deal of noise radiation. The higher the frequencies, the smaller the surface area must be to prevent disturbance. Since machines will always vibrate to some extent, noise control will be aided if the machines are kept as small as possible.

Principle

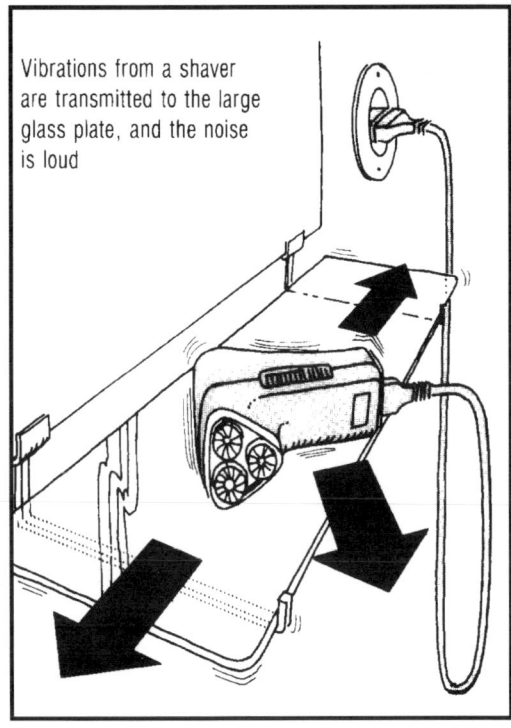
Vibrations from a shaver are transmitted to the large glass plate, and the noise is loud

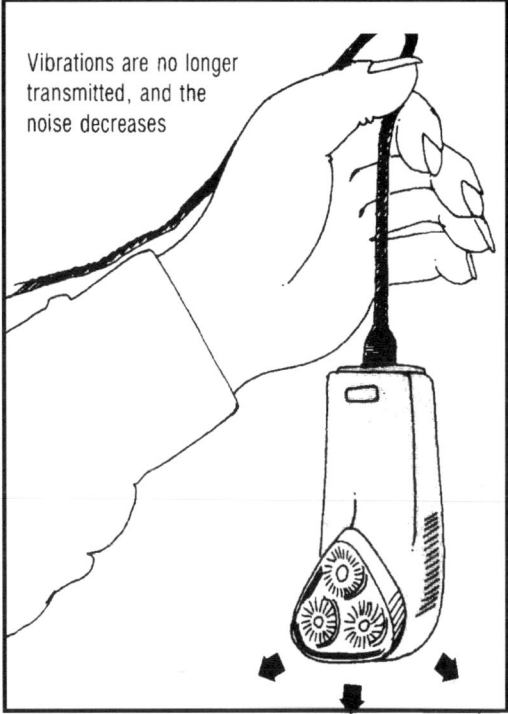
Vibrations are no longer transmitted, and the noise decreases

Example

Excessive noise is radiated from the control panel of a hydraulic system.

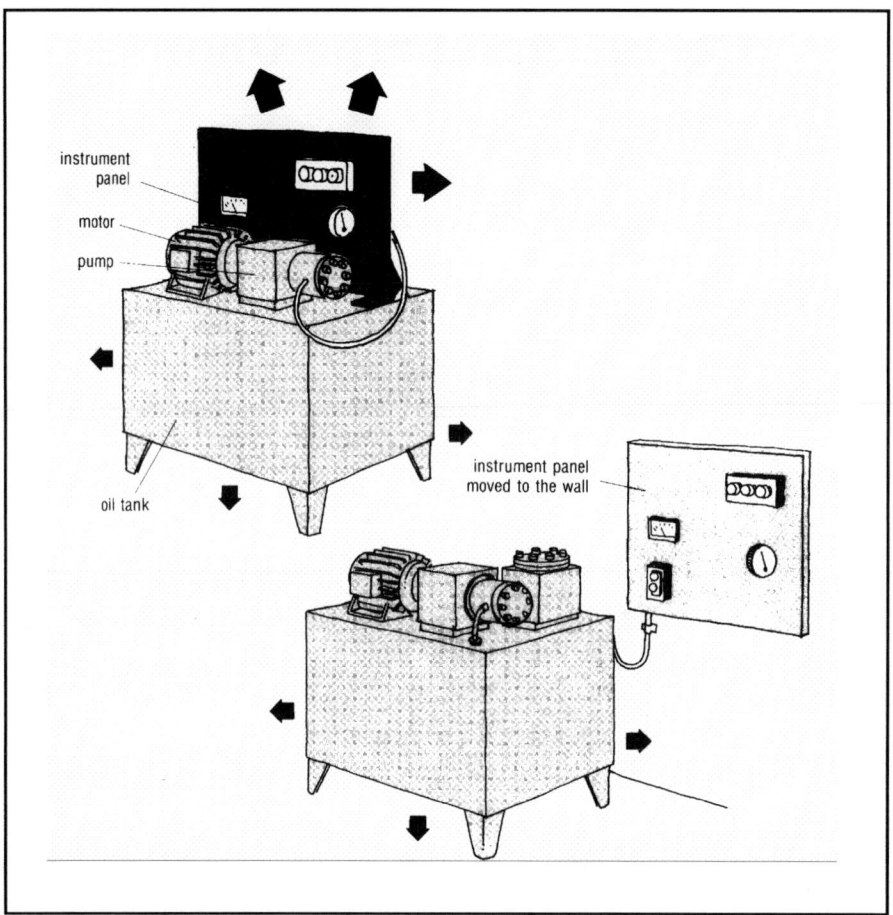

Control Measure

The panel is detached from the system itself, the vibrating surface is reduced, and therefore the noise level is decreased.

Densely Perforated Plates Produce Less Noise

Large vibrating surfaces cannot always be avoided. The vibrating surface pumps air back and forth like the piston of a pump, causing sound radiation. If the panel is perforated, the "piston" leaks, and the pumping functions poorly. Alternatives to perforated plated include mesh, gratings and expanded metal.

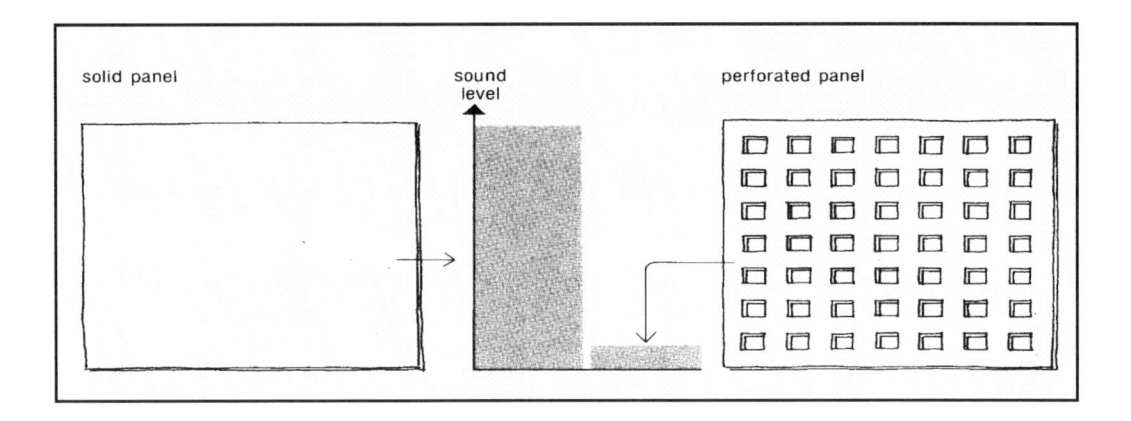

Principle

Example

The protective cover over the flywheel and belt drive of a press is a major noise source. The cover is made of solid metal.

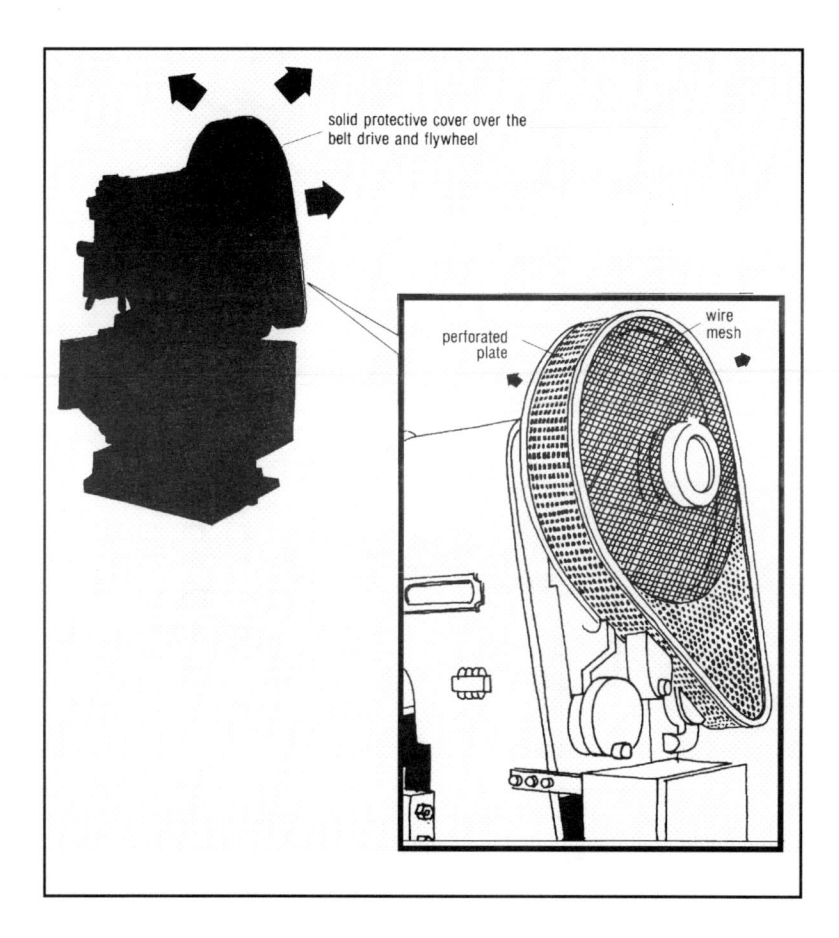

Control Measure

A new cover is made of perforated sheet metal and wire mesh. Sound radiation is reduced.

A Long, Narrow Plate Produces Less Sound Than a Square One

When a plate is set into vibration, excess air pressure forms on one side of the plate and then the other. Sound comes from both sides. The pressure difference balances out close to the edges so the noise radiation there is slight. Therefore, a long, narrow plate radiates less sound.

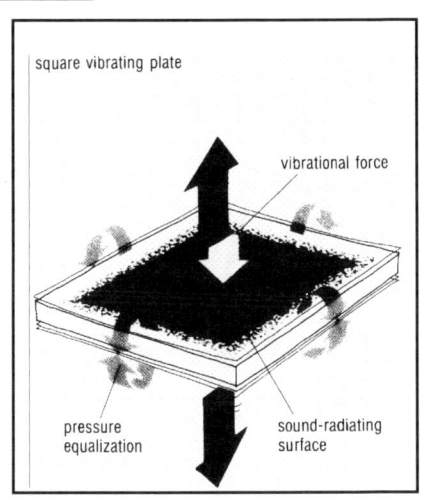

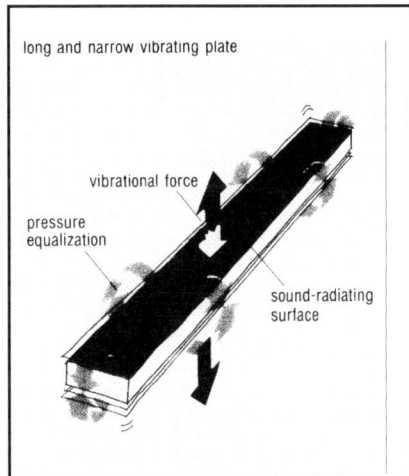

Principle

Application of Noise Control Principles

Example

A belt drive provides a large amount of low frequency noise because of the vibration of the broad belt.

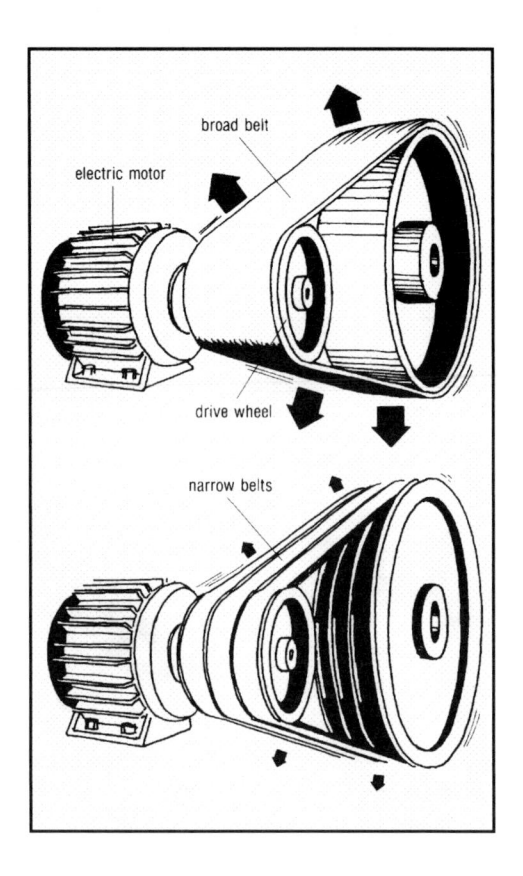

Control Measure

The broad belt is replaced by narrower belts, separated by spacers. This reduces the noise level.

Plates with Free Edges Produce Less Low-Frequency Noise

If a plate vibrates with free edges, pressure equalization takes place between the two sides of the plate, thus reducing sound emissions. Clamping the corners prevents pressure equalization and the sound emission is greater, especially at low frequencies. For example, speakers produce more bass sound if they are enclosed in a cabinet.

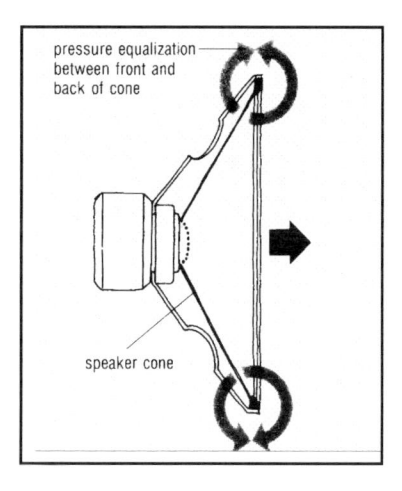

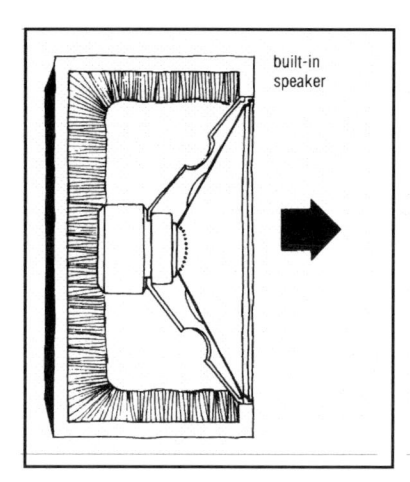

Principle

Example

Bumps in the floor produce noise from the bottom and side plates of a cart when the cart is pushed. Sound is also emitted when material is slide down the cart walls. Pressure equalization only takes place at the top edges of the size plates.

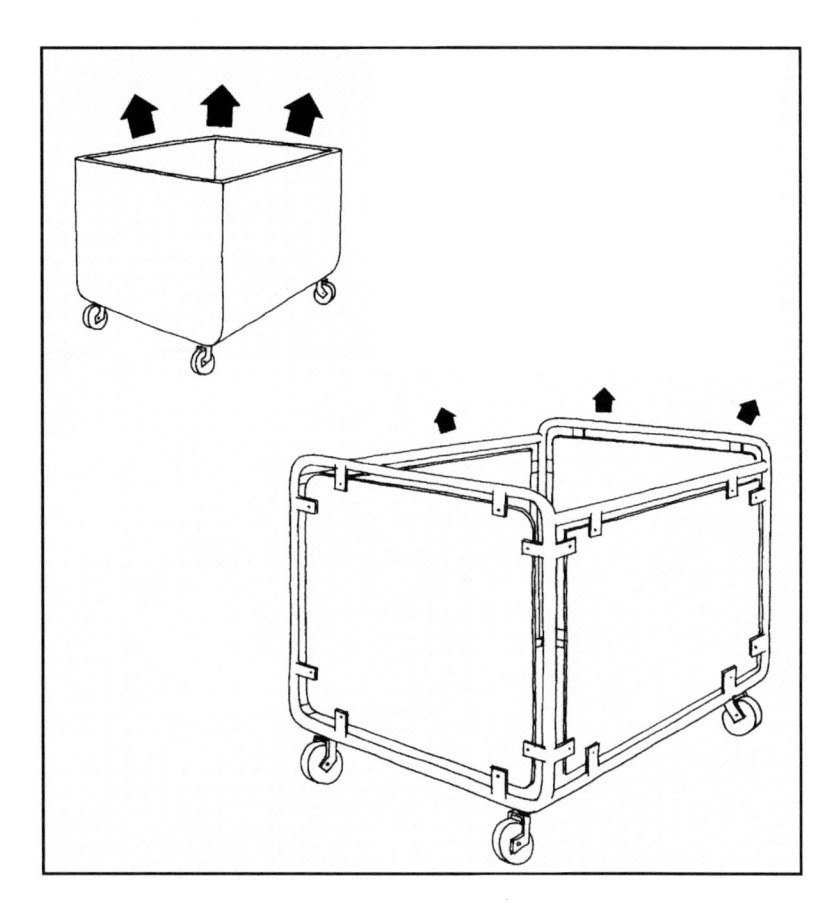

Control Measure

The walls are replaced by new ones, constructed with a pipe frame. Plates are fastened with a gap between the plates and the frame. Pressure equalization takes place along all the edges, and the low-frequency noise is reduced.

Light Objects and Low Speed Produce the Least Impact Noise

When a plate is struck by an object, the plate vibrates and makes noise. The sound level is determined by the weight of the object and its striking speed. If the ropping height of an object is reduced from 5 meters about 16 feet) to 5 centimeters (2 inches), the sound level drops about 20 dB.

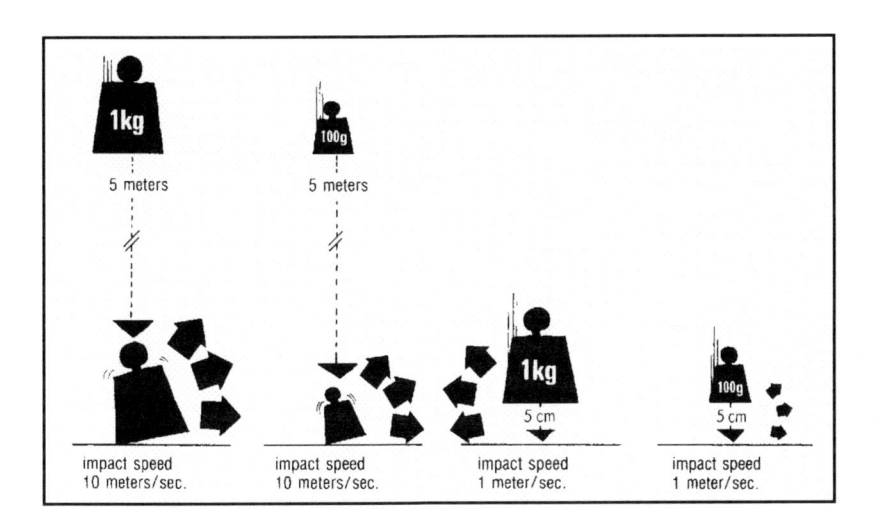

Principle

| 1kg, 5 meters, impact speed 10 meters/sec. | 100g, 5 meters, impact speed 10 meters/sec. | 1kg, 5 cm, impact speed 1 meter/sec. | 100g, 5 cm, impact speed 1 meter/sec. |

Example

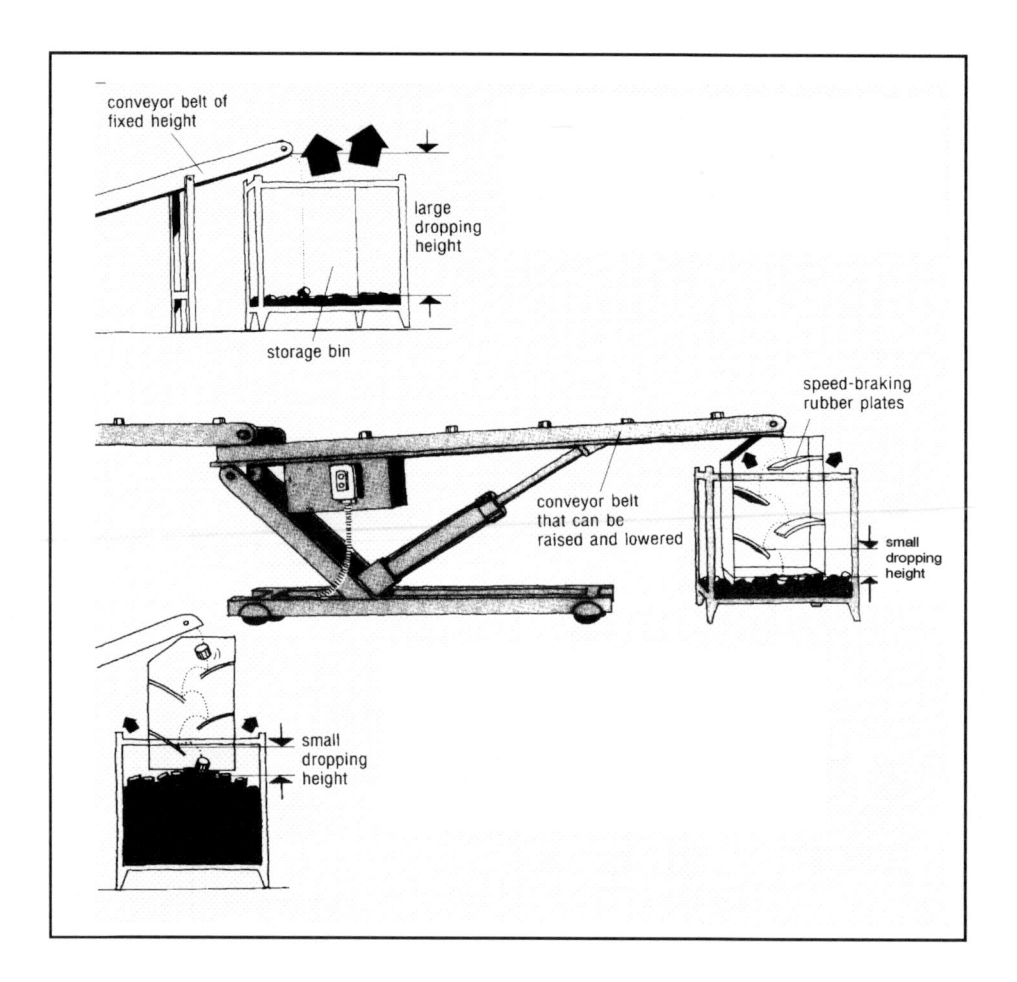

Control Measure

A hydraulic system is installed so that the conveyor belt can be raised and lowered. The belt ends in a drum equipped with rubber plates to break the fall of the parts. The drum is raised automatically.

A Damped Surface Gives Off Less Sound

As vibration moves throughout a plate, it gradually decreases as it travels, but in most plates this reduction is rather small. In such cases the material is said to have low internal damping. Internal damping in steel, for example, is extremely poor. Good damping can be achieved by adding coatings or intermediate layers with better internal damping than the base material.

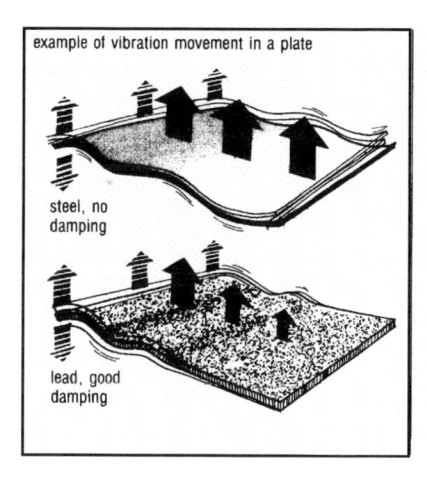

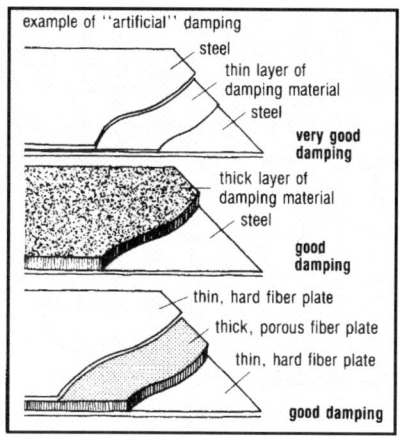

Principle

Example

The loudest noise from a pump system comes from the coupling guard which is made of sheet metal.

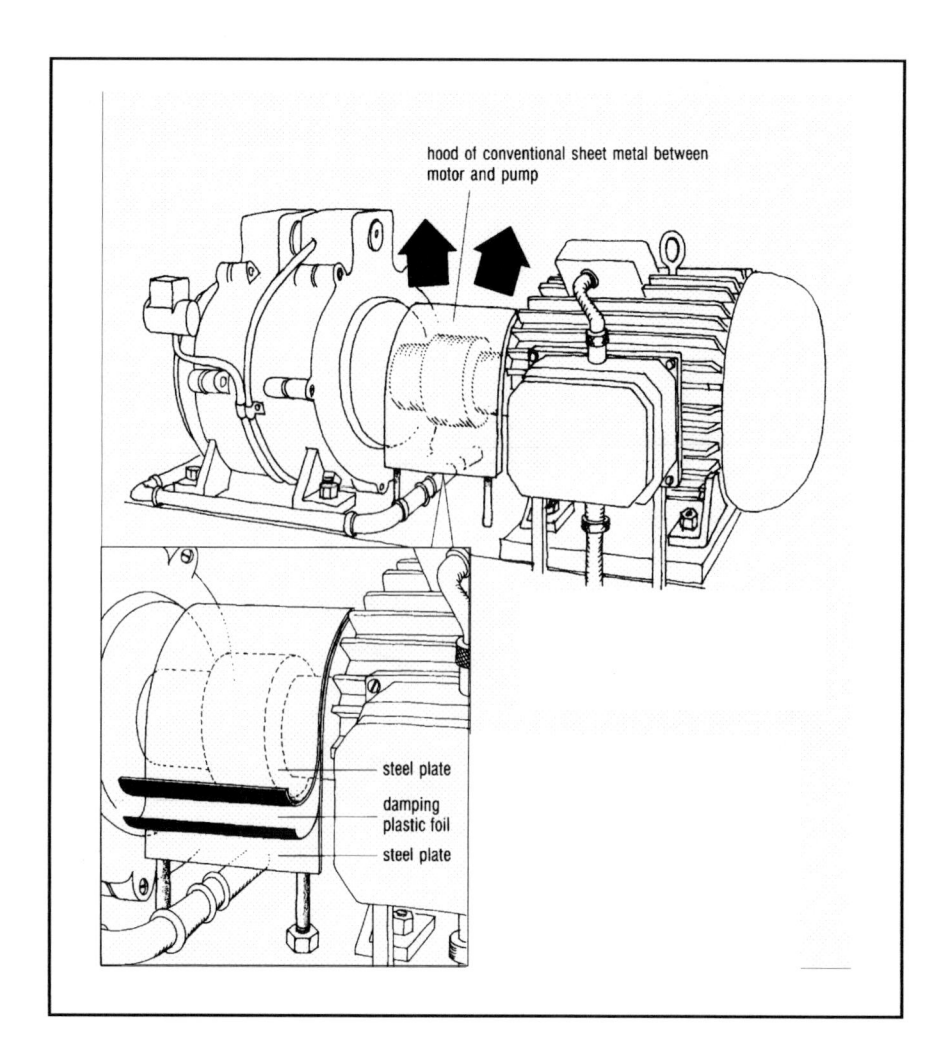

hood of conventional sheet metal between motor and pump

steel plate
damping plastic foil
steel plate

Control Measure

The noise level is reduced by isolating the vibration of the guard or constructing it of damped metal. If the coupling creates a siren-type noise, the guard may need an acoustical lining.

Resonance Increases Noise But It Can Be Damped

Resonance greatly increases noise from a vibrating plate, but it can be suppressed or prevented by damping the plate. It may often be sufficient to damp only part of the surface and, in some rare cases, damping of a single point is effective.

Principle

A tap against a glass produces a loud ring

Damping eliminates the ring

Example

An automatic tooth cutter for circular saw blades produces intense resonance sound.

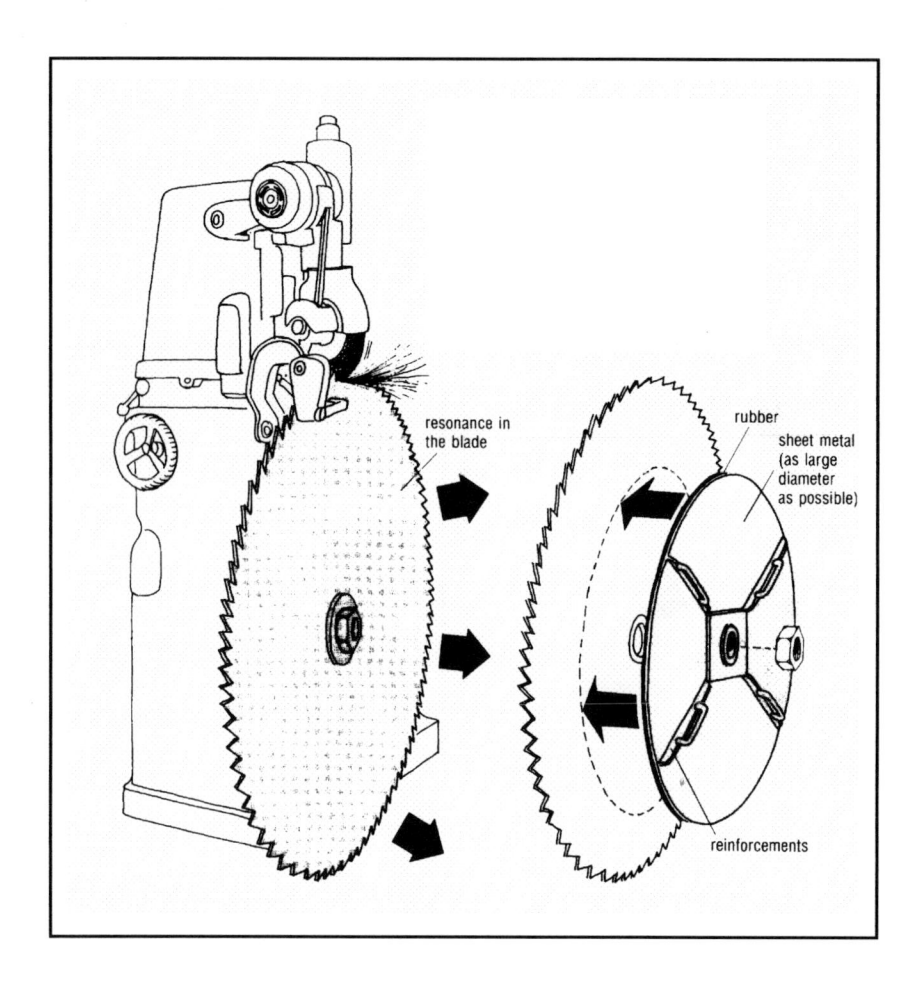

Control Measure

A urethane rubber coating clamped to the saw blade damps the resonance.

Resonance Shifted to Higher Frequency is More Easily Damped

Large vibrating plates often have low-frequency resonance which can be difficult to damp. If the plate is stiffened, the resonance shifts to higher frequency, which can be more easily damped.

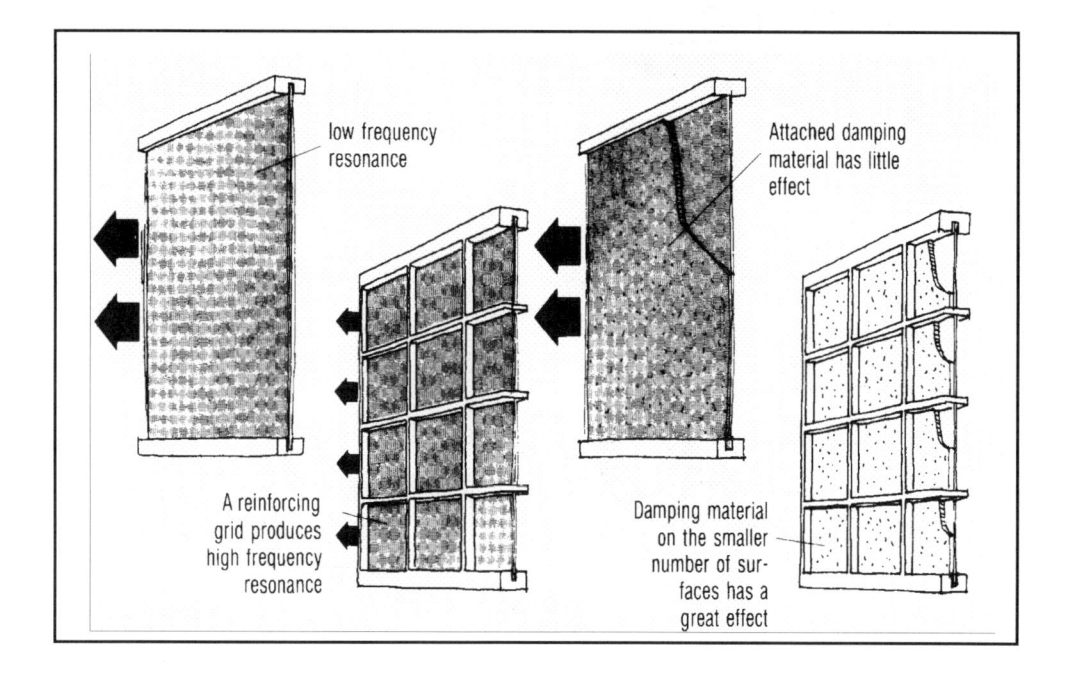

Principle

low frequency resonance

A reinforcing grid produces high frequency resonance

Attached damping material has little effect

Damping material on the smaller number of surfaces has a great effect

Application of Noise Control Principles

Example

The greatest low-frequency sound from this machine comes from the side surfaces of the machine stand.

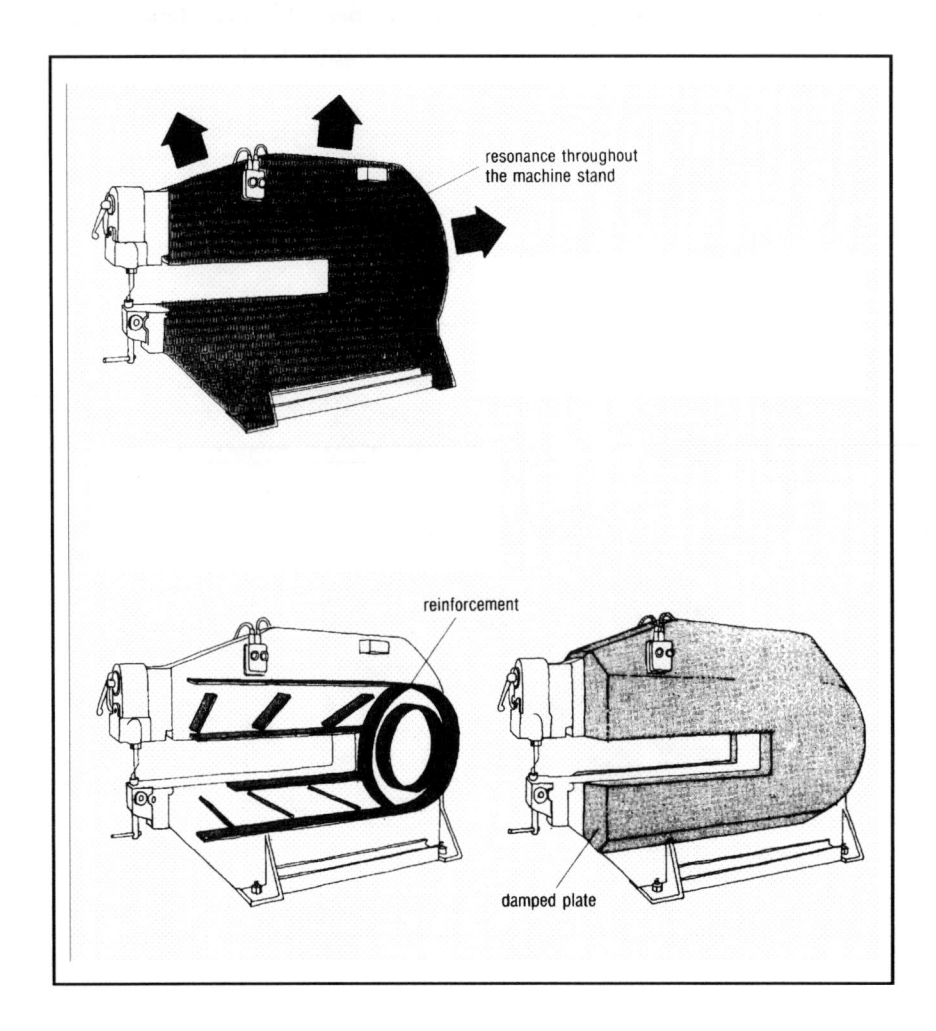

Control Measure

The side plates on the machine are stiffened with iron straps. A damped plate is installed over the braces.

Sound Production in Air or Gases

Wind Tones Can Be Eliminated

When air passes by an object at certain speeds, a strong pure tone, known as a Karman Tone, can be produced. This can be prevented by making the object longer in the direction of flow, such as with a "tail," or by making the object's shape irregular.

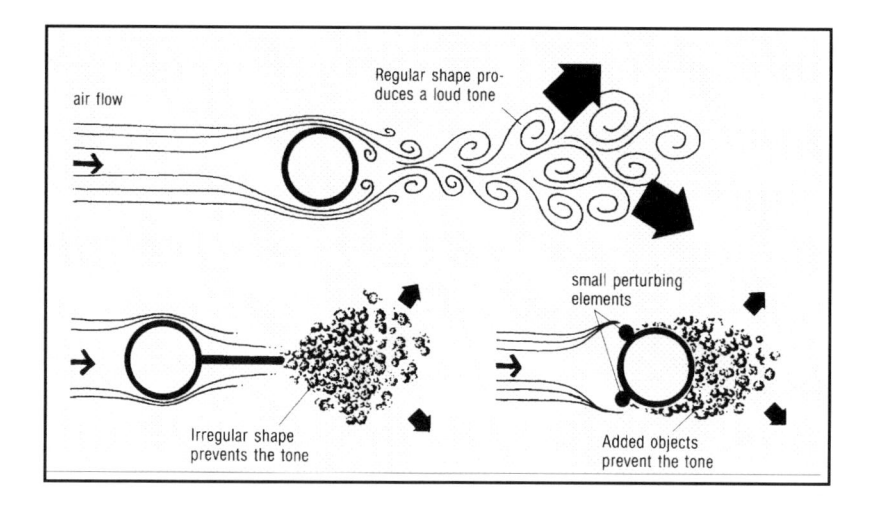

Principle

Example

At certain wind speeds, loud sounds can form around stacks, causing disturbances.

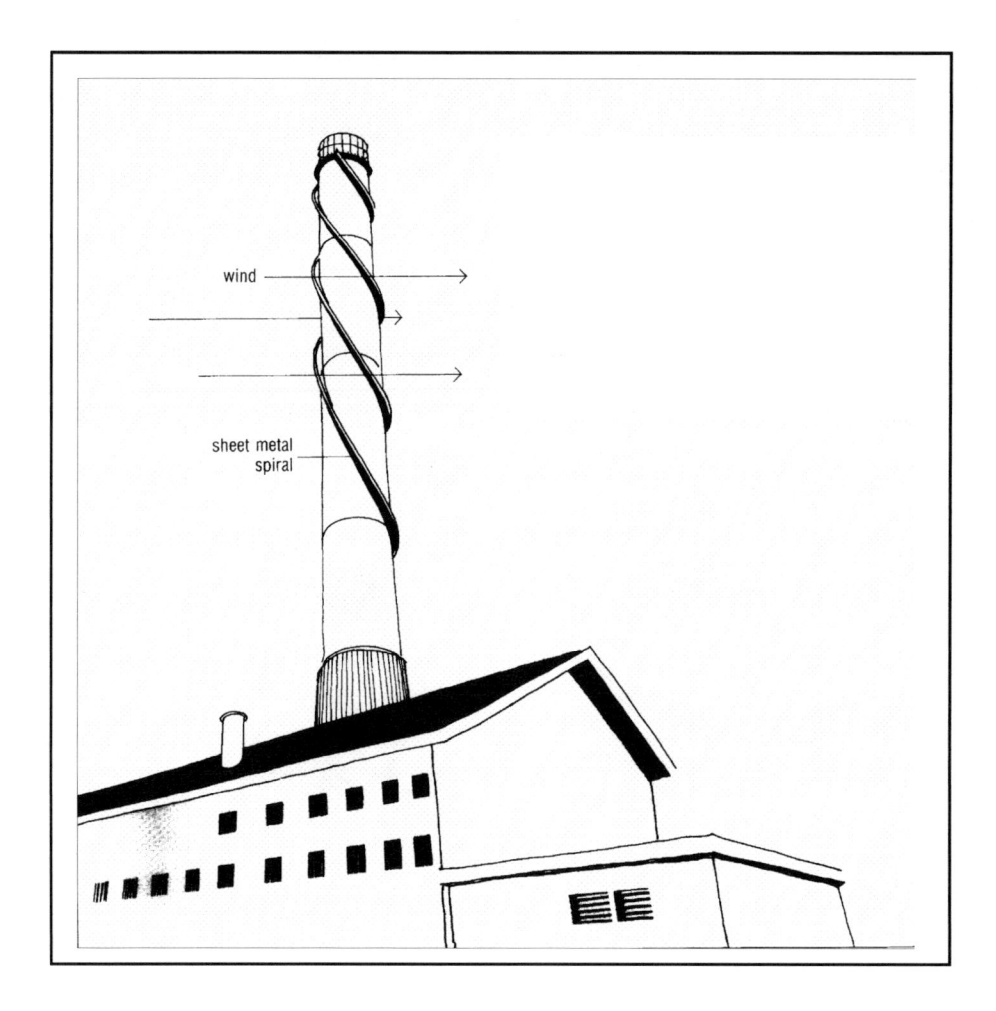

Control Measure

A strip of sheet metal is mounted on the stack in a spiral. The pitch of the spiral must *not* be constant. Regardless of the wind's direction, it encounters an irregular object.

Air Flow Past Hollow Openings Should Be Avoided

When air or another gas blows across the edges of an opening to a hole, loud, pure tones are formed. This is how a wind instrument operates. The greater the volume of the hole and the smaller the number of openings, the lower the frequency of the tone.

Principle

Example

When a cutter revolves under no-load conditions, sound can arise from the track for holding the plane blade. An air stream is being chopped, creating a siren (pure tone) noise.

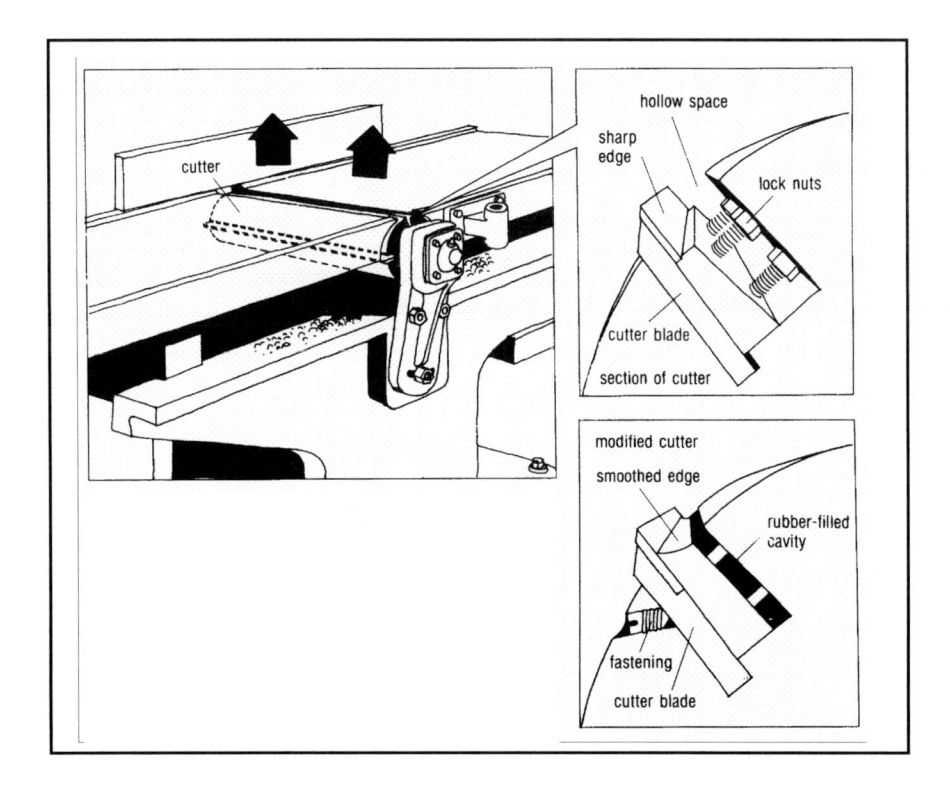

Control Measure

Minimizing the cavities by filling the empty space in the track with a rubber plate reduces the pumping action and the noise.

Pipes Without Impediments Produce the Least Amount of Noise from Turbulence

During flow in pipes (or ducts) there is always some turbulence against the walls. The noise from turbulence is increased if the flow must rapidly change direction, if the flow moves at a fast rate, and if objects blocking the flow are close together.

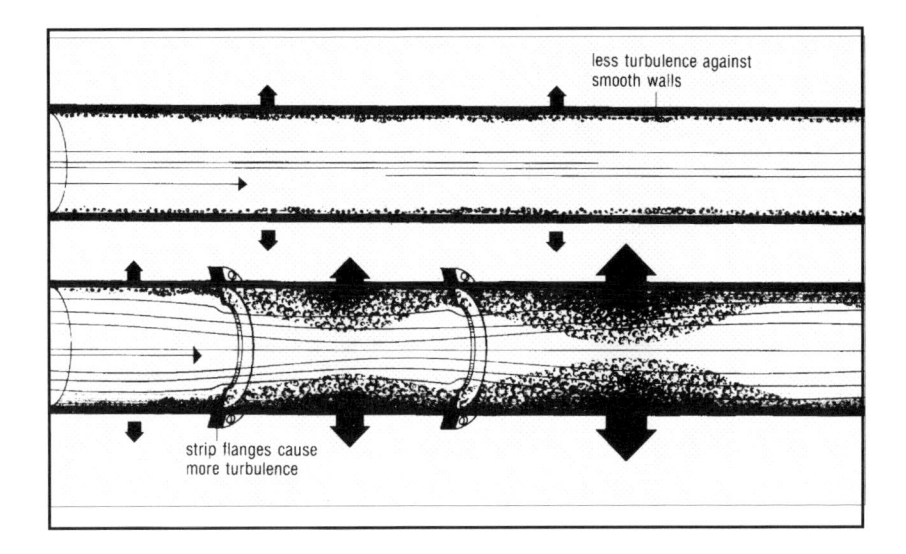

Principle

less turbulence against smooth walls

strip flanges cause more turbulence

Example

A branch of a steam line has three valves which produce a loud shrieking sound. The branch has two sharp bends which also produce a lot of noise.

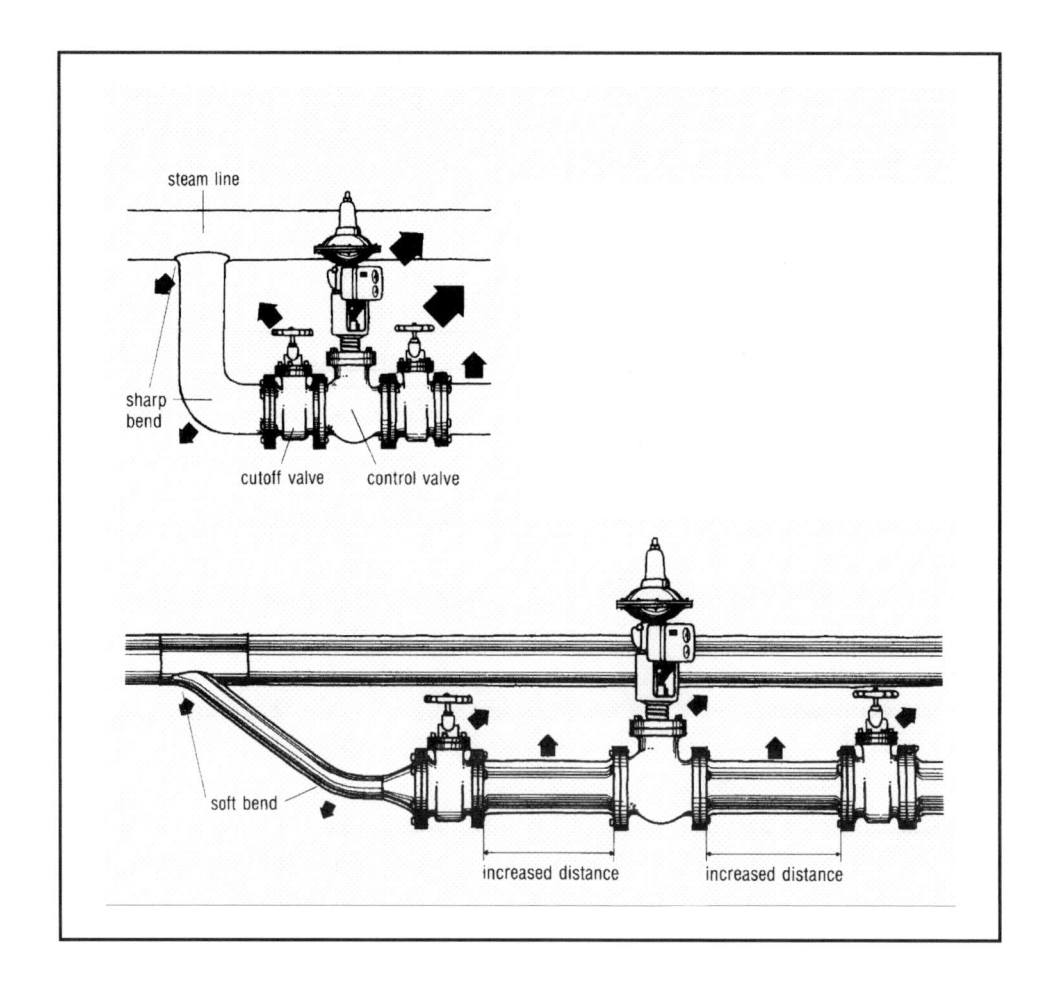

Control Measure

A new branch is created with softer bends. Tubing pieces are placed between the valves so that turbulence will be reduced or eliminated before the steam reaches the next valve.

Undisturbed Flow Produces the Smallest Amount of Exit Noise

When a flowing gas mixes with non-moving gas, noise may be produced, especially if the flow is disturbed before the outlet. A lower outflow speed will produce a lower sound level. For speeds below 325 feet/sec., reduction of the speed by half will mean that the sound will be reduced by about 15 dB.

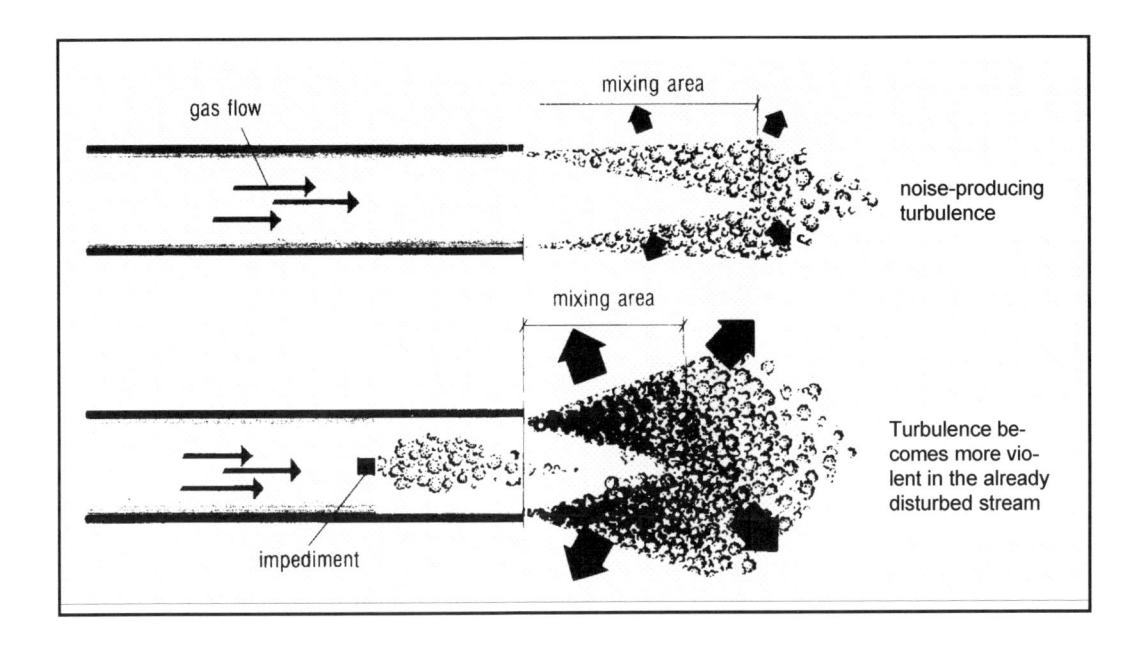

Principle

Example

The exhaust air from a compressed air-driven grinding machine produced a loud noise. The air becomes turbulent while leaving the machine through the side handle.

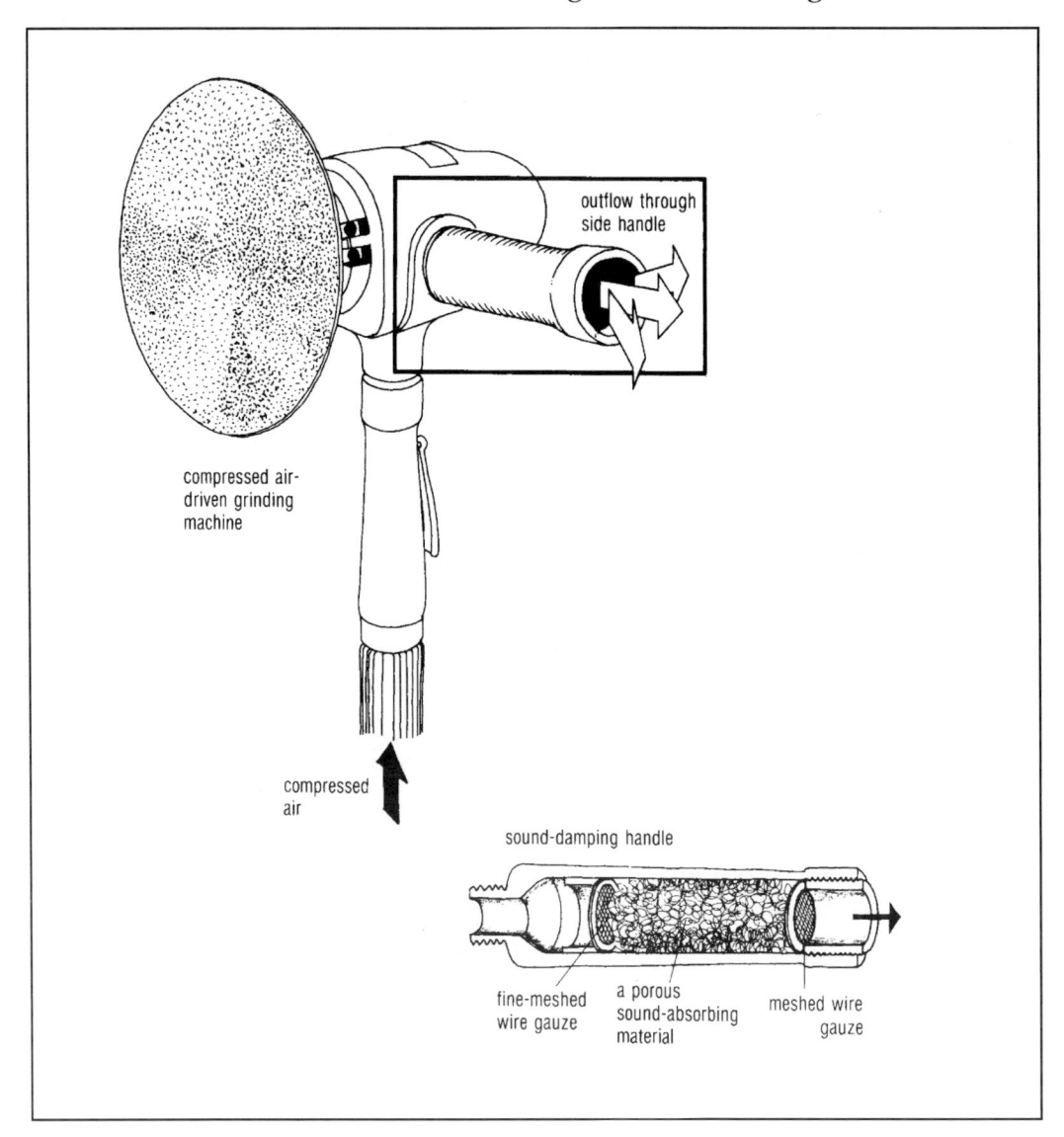

Control Measure

 A new handle is developed, filled with a porous sound-absorbing material between two fine-meshed gauzes. Passage through the porous materials breaks up the turbulence. The air stream leaving the handle is less disturbed, and the exhaust noise is weaker. A straight lined duct-type muffler may also be used.

Jet Noise Can Be Reduced by Using an Extra Air Stream

The term "jet stream" applies at flow speeds in excess of 325 ft/sec. Turbulence outside the outlet is great. Reducing the outflow speed by half may decrease the noise level by as much as 20 dB. Since the noise level is determined by the speed of the jet stream in relation to the speed of the surrounding air, noise production can be greatly reduced by using an air stream with a lower speed outside the jet stream.

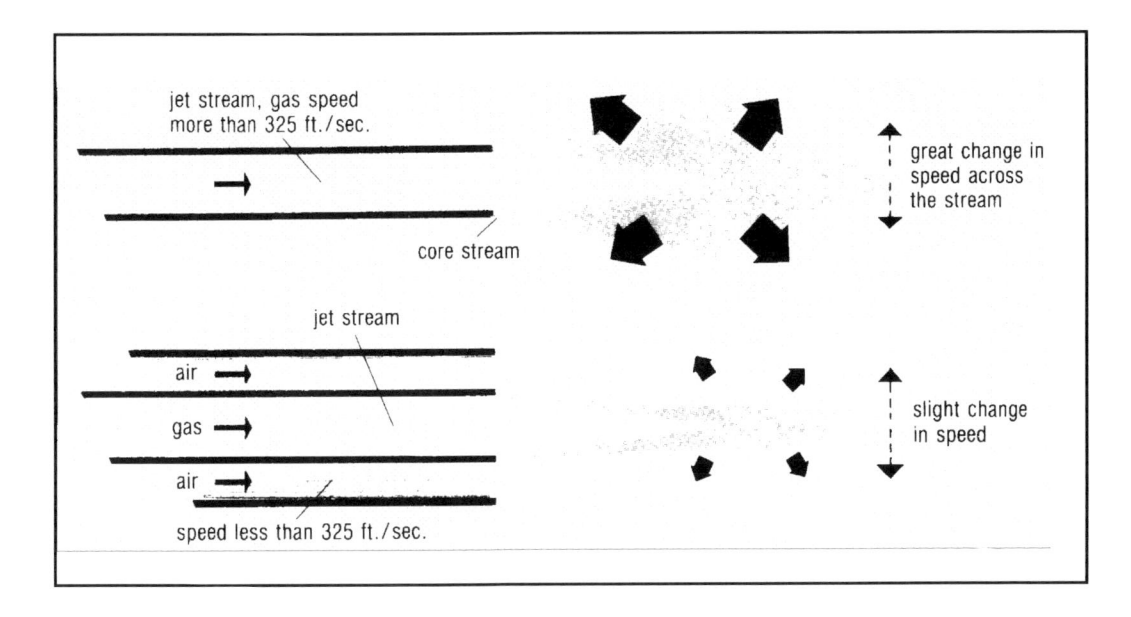

Principle

Example

The cleaning of machine parts with compressed air after processing is often carried out with simple tubular mouthpieces. Very high exit speeds are required, and a strong high-frequency noise develops.

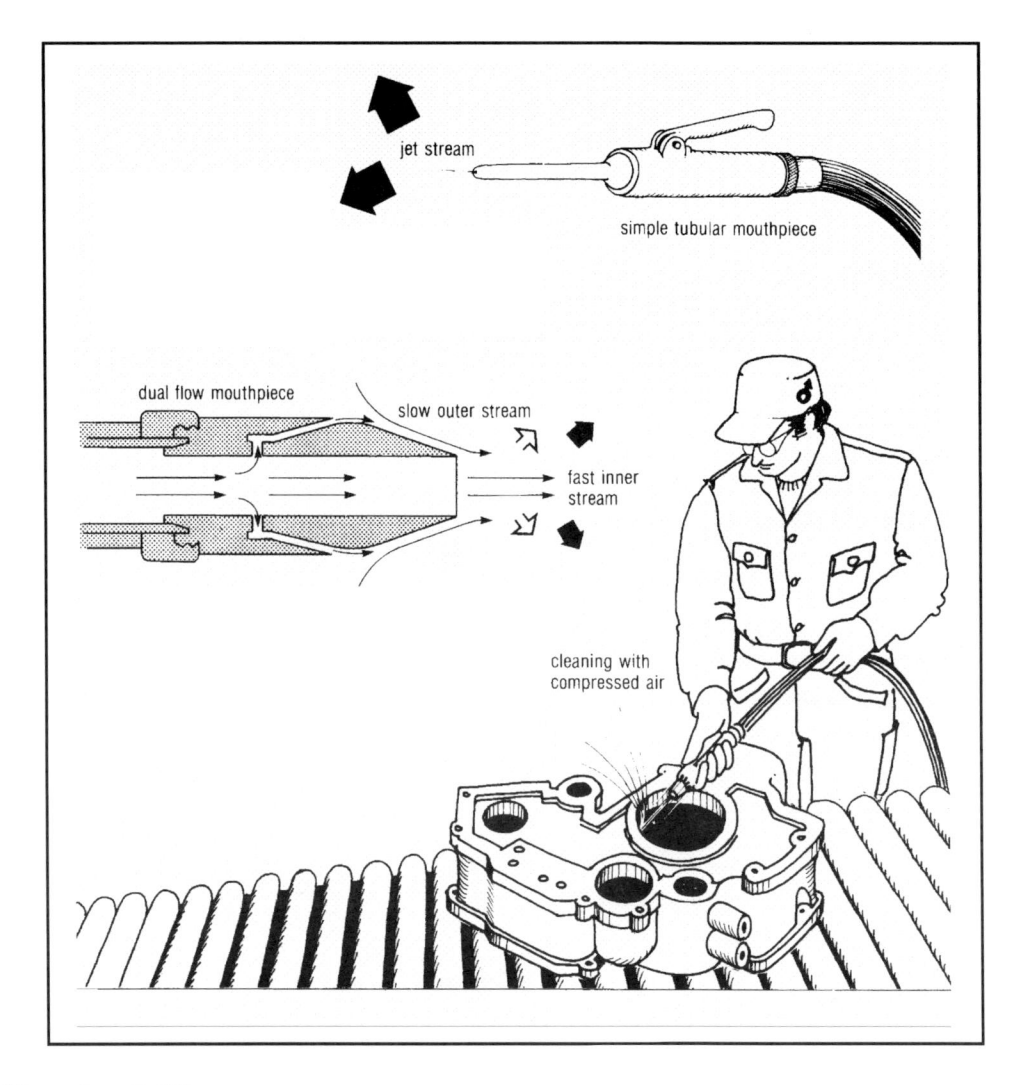

Control Measure

The simple tubular mouthpiece can be replaced by mouthpieces which produce less noise, such as a dual flow mouthpiece. In this mouthpiece, part of the compressed air moves at a lower speed outside the central stream.

Low Frequency Jet Noise is Easier to Reduce if Converted to High Frequency

If the diameter of a gas outlet is large, the noise will peak at the low frequency. If the diameter is small, the noise will peak at high frequency. The low-frequency noise can be reduced by replacing a large outlet with several small ones. To some extent this will increase the high-frequency noise, but this is more easily controlled.

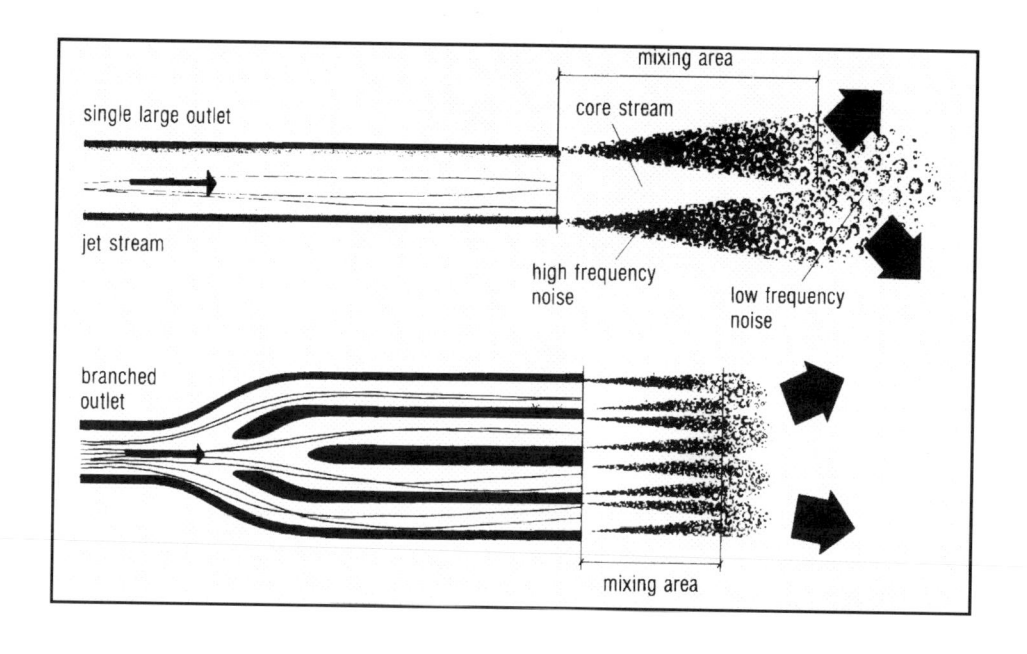

Principle

Application of Noise Control Principles

Example

Steam safety valves may discharge many times each day. Sound production during steam escape can produce high-level, low-frequency sound.

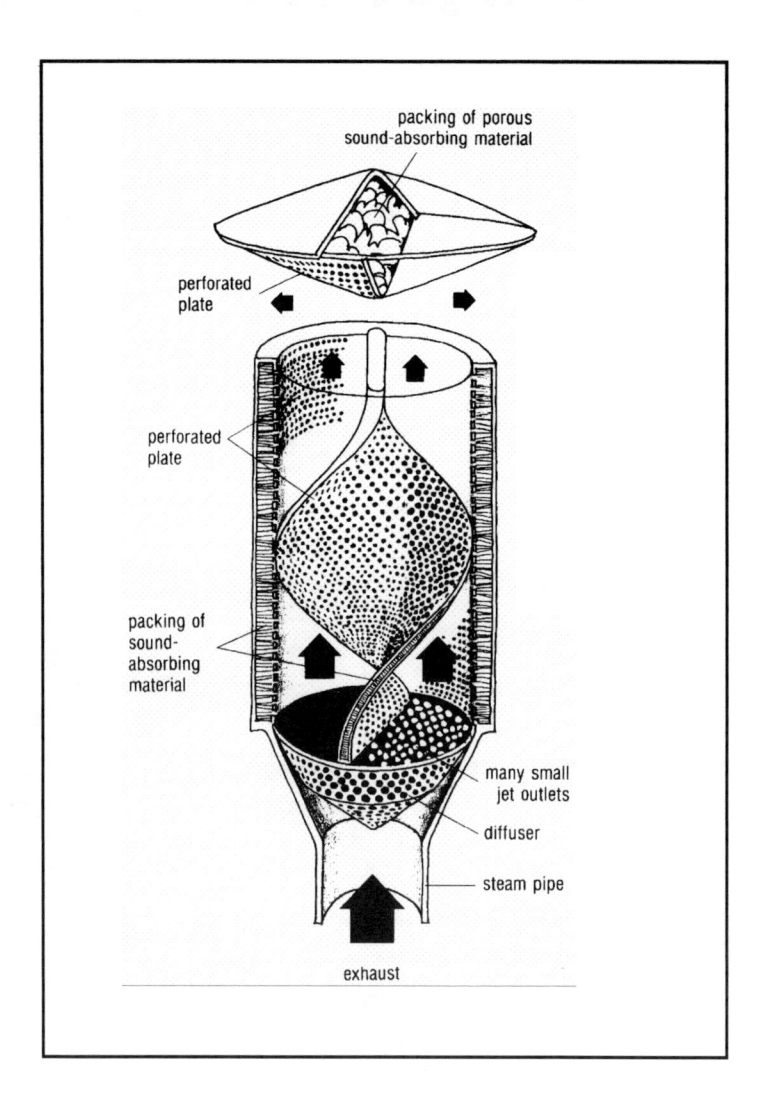

Control Measure

A diffuser is formed as a perforated cone. The holes produce many small jet streams and high-frequency noise which is absorbed in the downstream pack.

Fans Make Less Noise if Placed in Smooth, Undisturbed Flow Streams

A fan produces turbulence in air, which causes noise. If turbulence is already present in the incoming air, the sound will be more intense. The same principle applies, for example, to propellers in water.

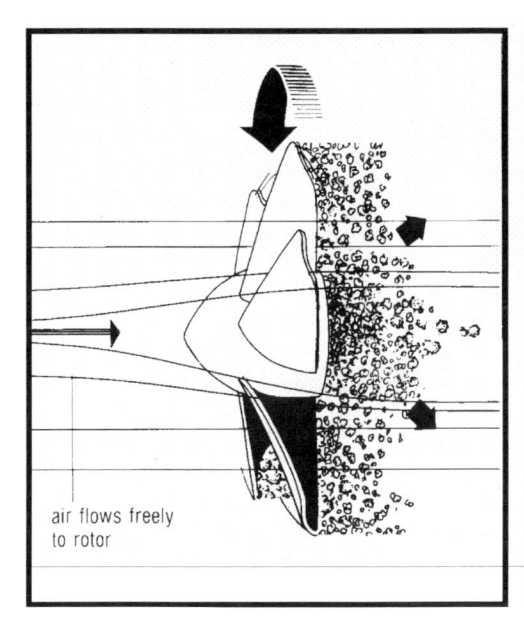

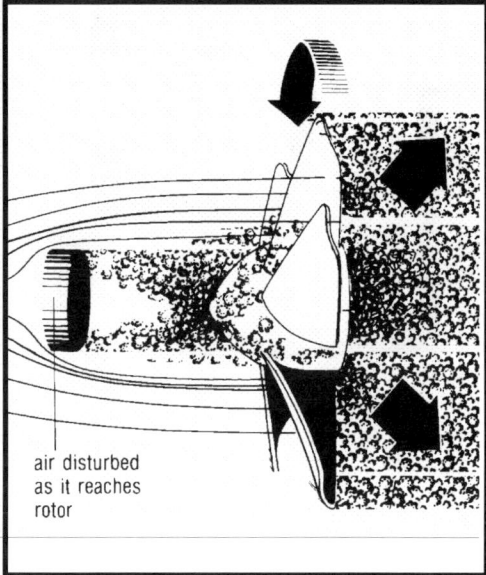

Principle

air flows freely to rotor

air disturbed as it reaches rotor

Example

In one case, the fan is located too close to a barrier, and in the other case, too close to a sharp bend. The flow is disturbed and the noise at the outlet is intense.

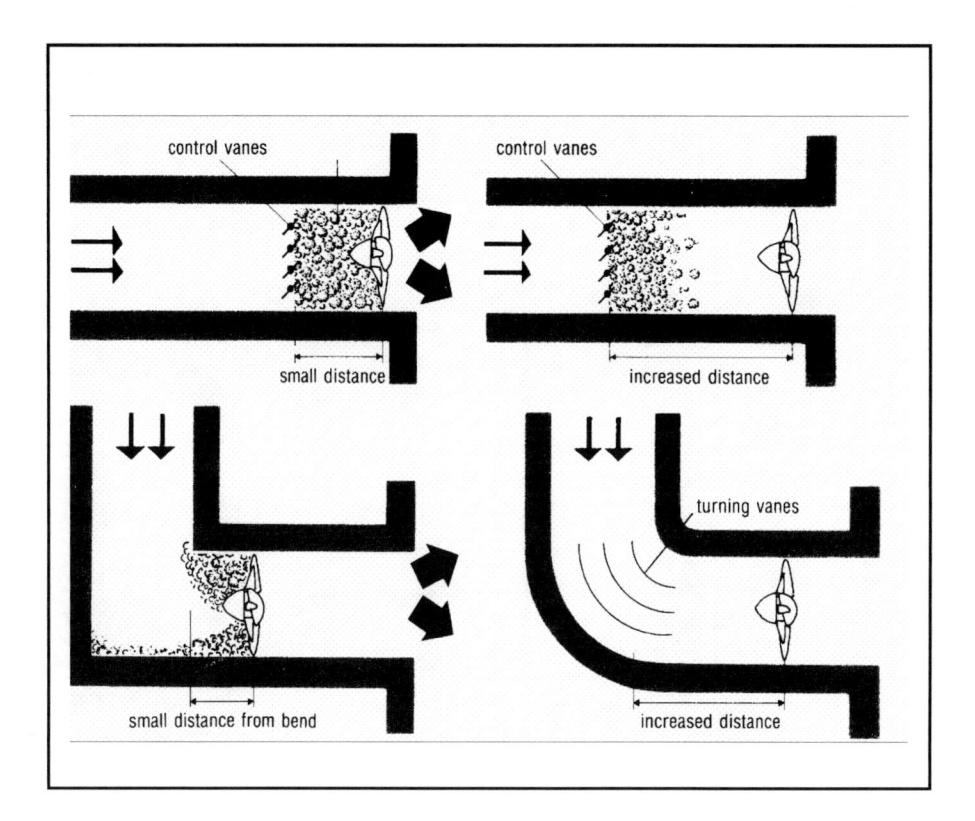

Control Measure

The control vanes are moved farther from the fan so that the turbulence has time to subside. In the other case, the bend is made smoother, and the fan is moved away from the bend. Turning vanes could also be used.

Sound Production in Flowing Liquids

Rapid Pressure Changes Produce More Noise

Turbulence will form if the pressure in a liquid system drops rapidly. Gas is released in the form of bubbles and produces a roaring noise. The pressure drop can be produced by a large, rapid change in volume. Noise is avoided by a gradual change in volume.

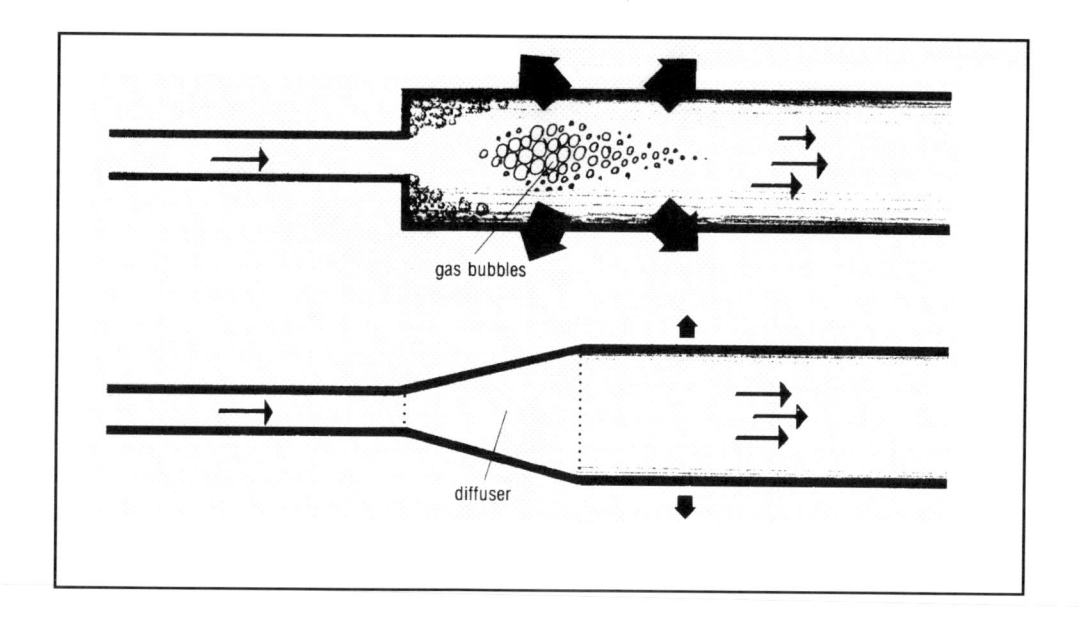

Principle

Example

Control valves in liquid systems often have small valve seats, resulting in large flow speeds with large pressure changes. Twisted flow pathways and sharp edges produce intense turbulence. Sound radiates directly from valves and pipes, and solid sound is conducted to walls.

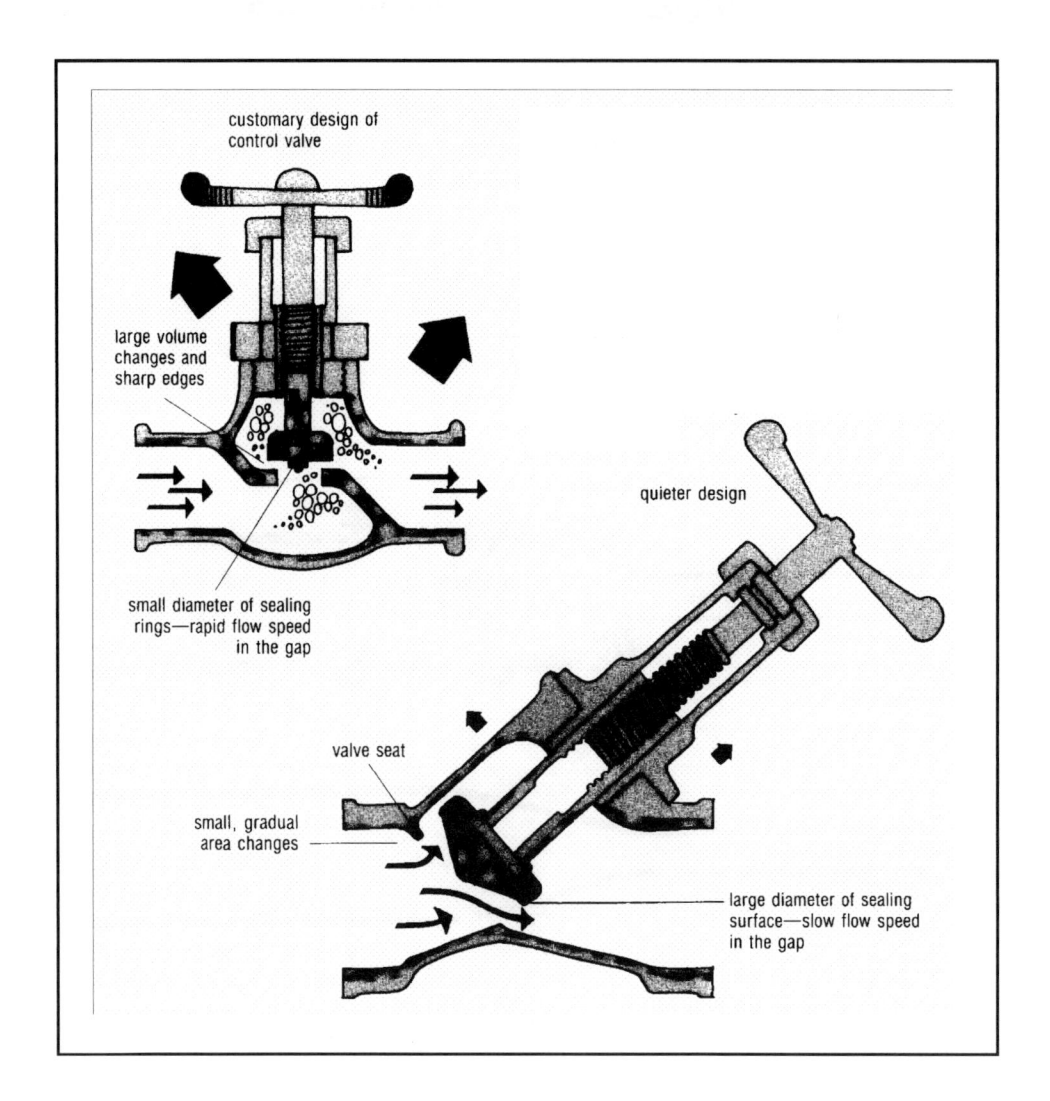

Control Measure

Control valves with larger cone diameters, straighter flow pathways, and more rounded edges are used.

Large and Rapid Changes in Pressure Produce "Cavitation" Sound

Noise production takes place at control valves, at pump pistons, and at propellers whn large and rapid pressure drops occur in liquids. This so-called "cavitation" noise is most common in hydraulic systems. Cavitation can be reduced by bringing about the pressure reduction in several smaller steps.

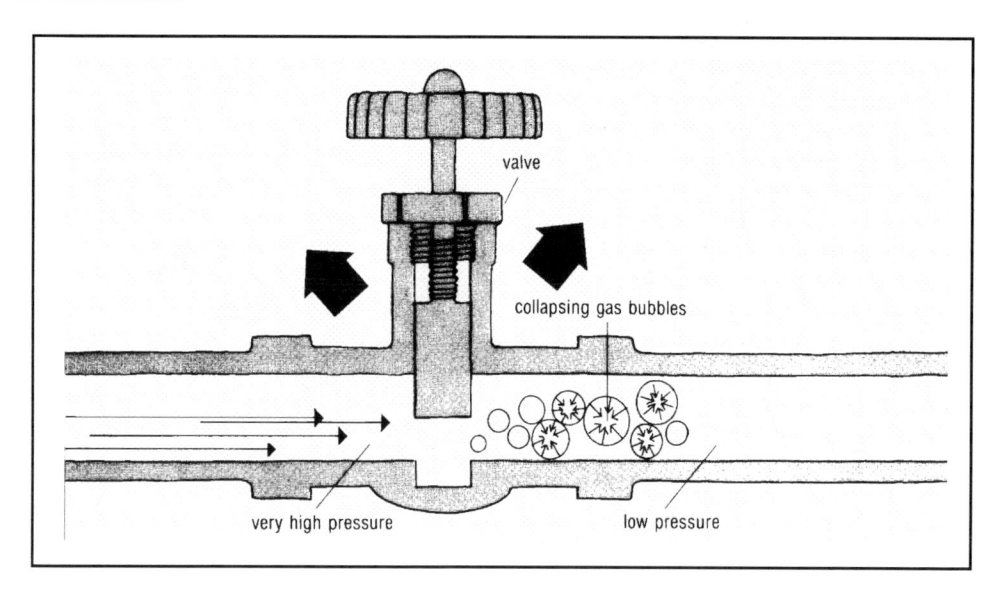

Principle

Example

In a hydraulic system, the full pump capacity is employed only in exceptional cases. The pressure is generally greatly reduced using a control valve. Cavitation can then arise, producing loud noise from the valve. The noise is conducted as solid-borne sound to connected machines and building structures.

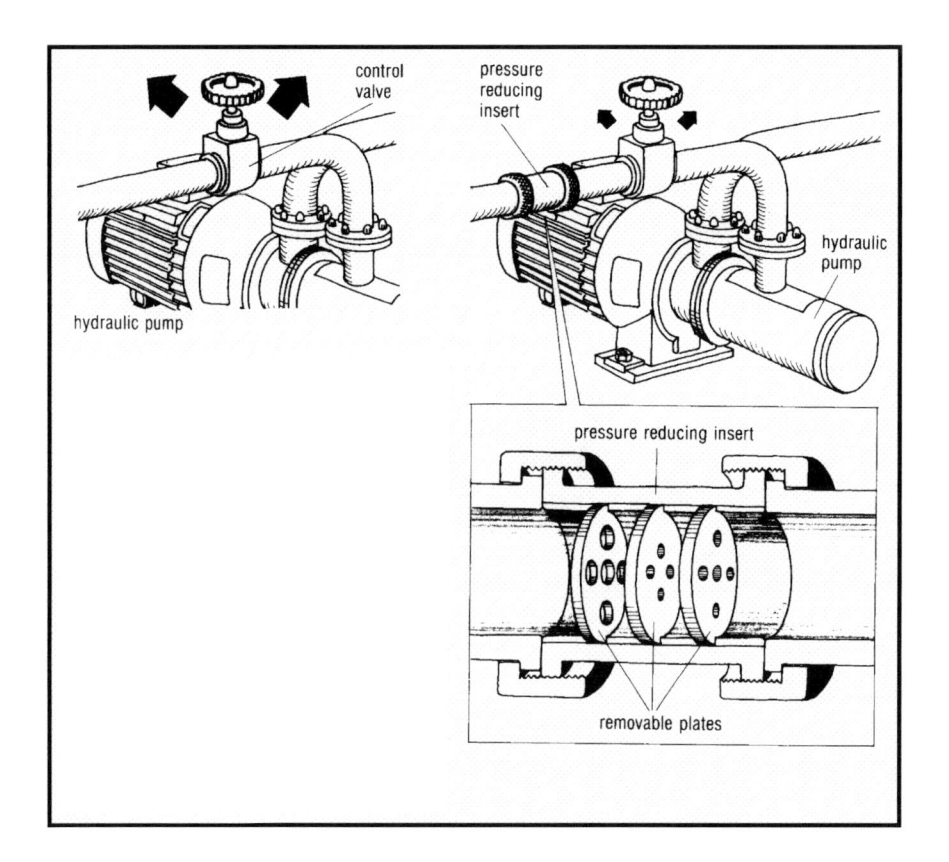

Control Measure

A pressure reducing insert is placed in the same pipe as the control valve. The insert has removable plates with different perforations. The plates are selected so that the insert will not produce a greater pressure drop than that required to prevent cavitation.

Sound Movement Indoors

Thick, Porous Layers Absorb Both High- and Low- Frequency Sound

Porous material through which air can be pressed often makes a good sound absorbent. Examples of such materials include felt, foam rubber, foamed plastic, textile fibers, and a number of sintered metals and ceramic materials. If the pores are closed, the absorption is slight. Thin, porous absorbents handle high tones. For good effects below 100 Hz, the thicknesses required may become impractical. Low-frequency absorption is improved with the aid of an air gap behind the absorbent.

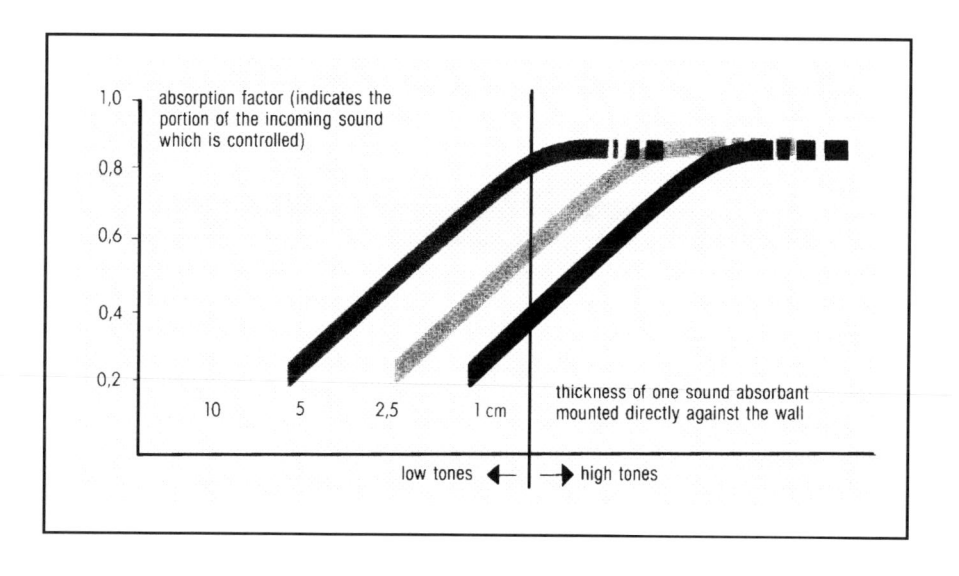

Principle

absorption factor (indicates the portion of the incoming sound which is controlled)

thickness of one sound absorbant mounted directly against the wall

low tones ← | → high tones

Application of Noise Control Principles

Example

A workshop with intense low-frequency noise is provided with absorbents that are effective for low tones. One part of the shop contains space for hanging absorption baffles which provide good low-frequency absorption and are easily installed. A traverse leaves no room for baffles in the other part of the shop. Instead, horizontal absorbent panels are installed above the traverse, 8 inches from the ceiling, to improve the low-frequency absorption.

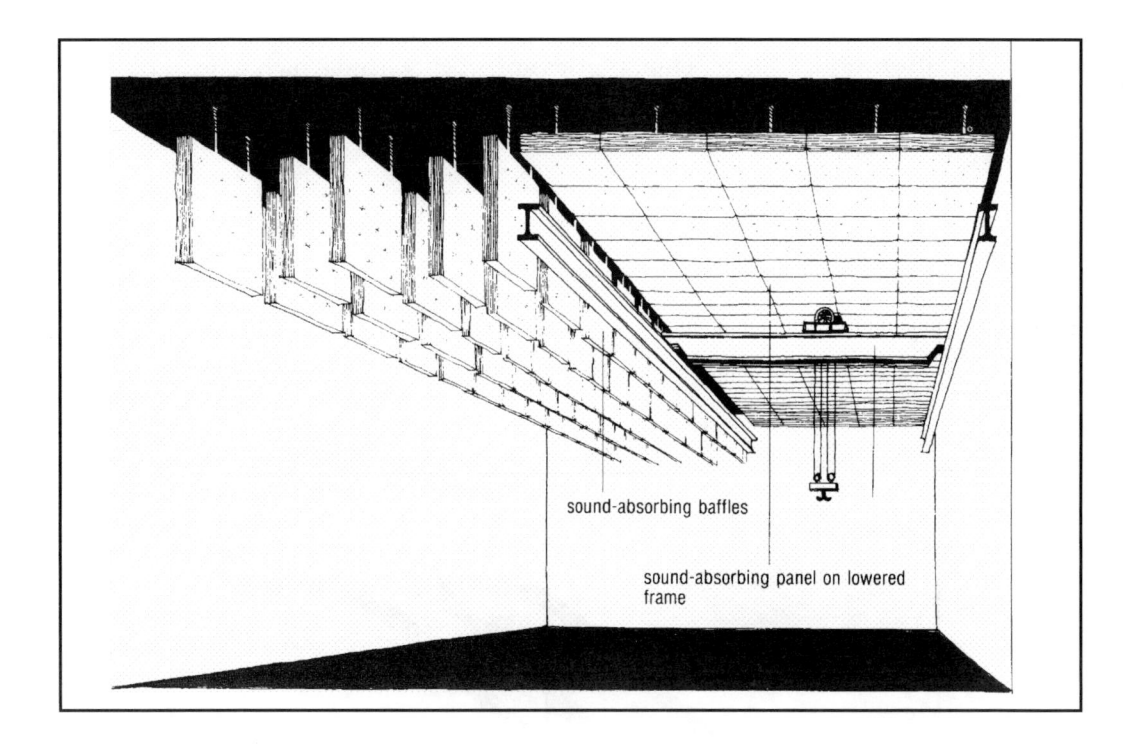

sound-absorbing baffles

sound-absorbing panel on lowered frame

Cover Layers with Large Perforations May Be Used Without Reducing Absorption

For a variety of reasons, a covering material may be needed to protect a porous absorbent. This can be done without reducing the effectiveness of the absorbent if the cover material has a sufficient number of openings. The thicker the cover layer, the larger the number of perforations that will be required.

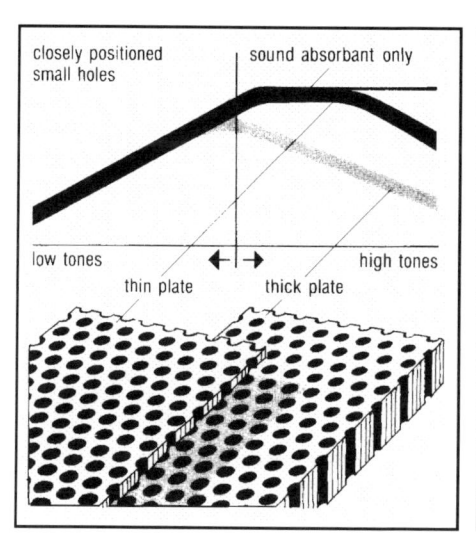

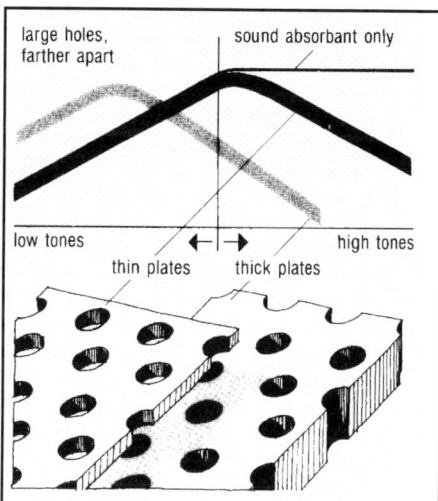

Principle

Example

Sound-absorbing material is required on many wall and ceiling surfaces in a building. To provide a more attractive environment, it is desirable to have many absorbents with different appearances.

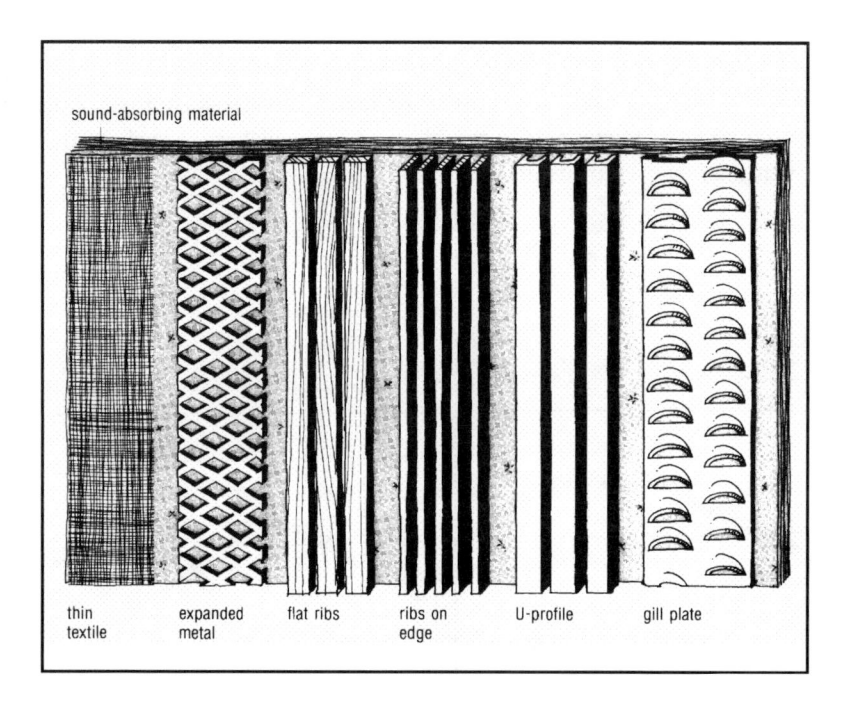

Control Measure

One material is used on all surfaces, with varying thicknesses. Different covering materials provide the desired variation in appearance.

Panels on Studs Absorb Low Frequencies

Thin panels, fastened to a system of studs, absorb low frequencies. The absorption is effective in a narrow frequency range. This range is determined by the stiffness of the panels and the distance between the fastenings. If the panels are fastened to studs on a wall, the distance from the wall also has an effect. A panel with large internal dampening absorbs in a larger frequency range. If a porous absorbing material is ued at these low frequencies, it must be very thick.

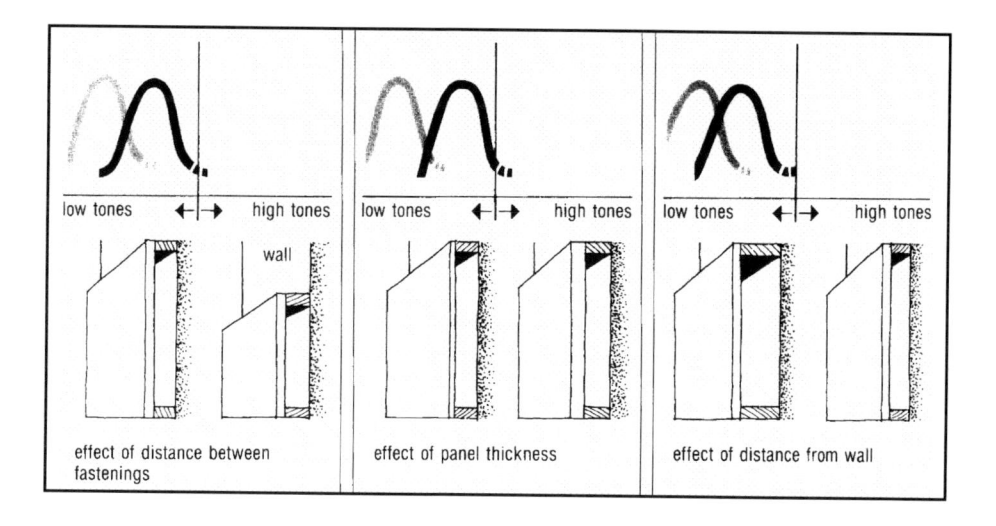

Principle

low tones ←→ high tones	low tones ←→ high tones	low tones ←→ high tones
effect of distance between fastenings	effect of panel thickness	effect of distance from wall

Application of Noise Control Principles

Example

Low-frequency resonance in an engine room produced a very loud hum near the walls and in the center of the room. When the revolution speed was significantly changed, the hum disappeared completely.

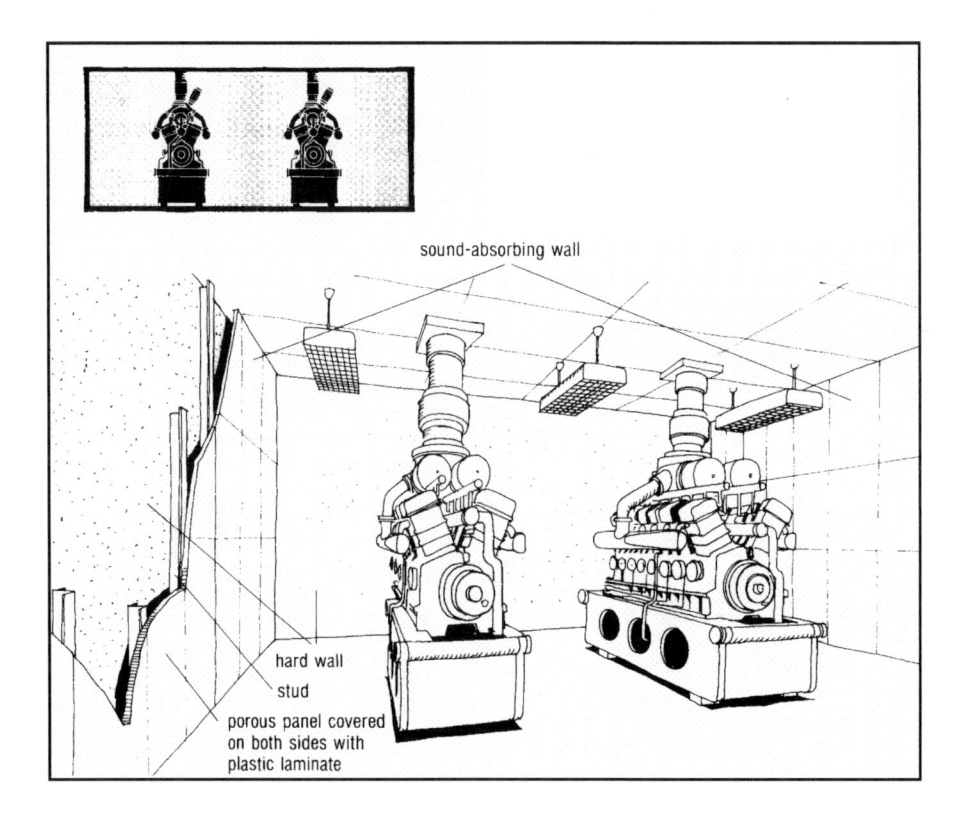

Control Measure

The walls were coated with panels on studs to provide the greatest absorption in the range of the loudest tone. In order for the absorbent to continue to function even in the case of slight deviations from the normal rotation speed, a layer with good internal damping was used, which provided a more extensive range with good absorption. As a result, the resonance and the loud hum disappeared.

Sound Shields May Be Combined with Sound Absorbing Ceilings

High-frequency noise can be reduced by using a shield. The shield is more effective the taller it is and the closer it is placed to the source. The effect of a shield is considerably reduced if the ceiling is not sound absorbent.

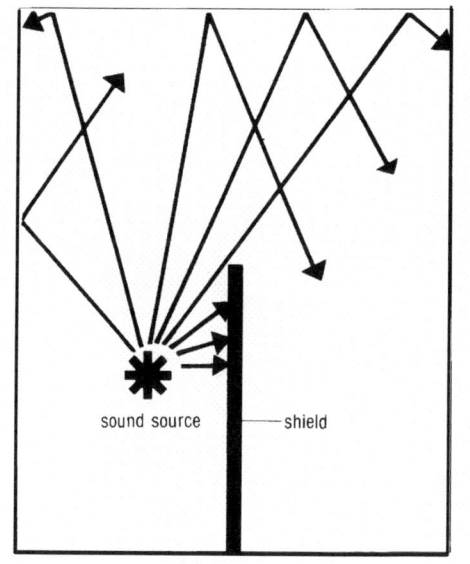

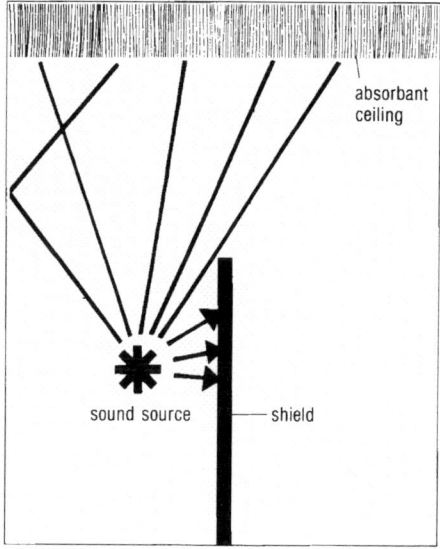

Principle

Example

In an auto plant with several assembly lines, the work on one line is noisier than the other. Grinding work on the bodies produces a shrieking, high-freuency sound, disturbing everyone in the plant.

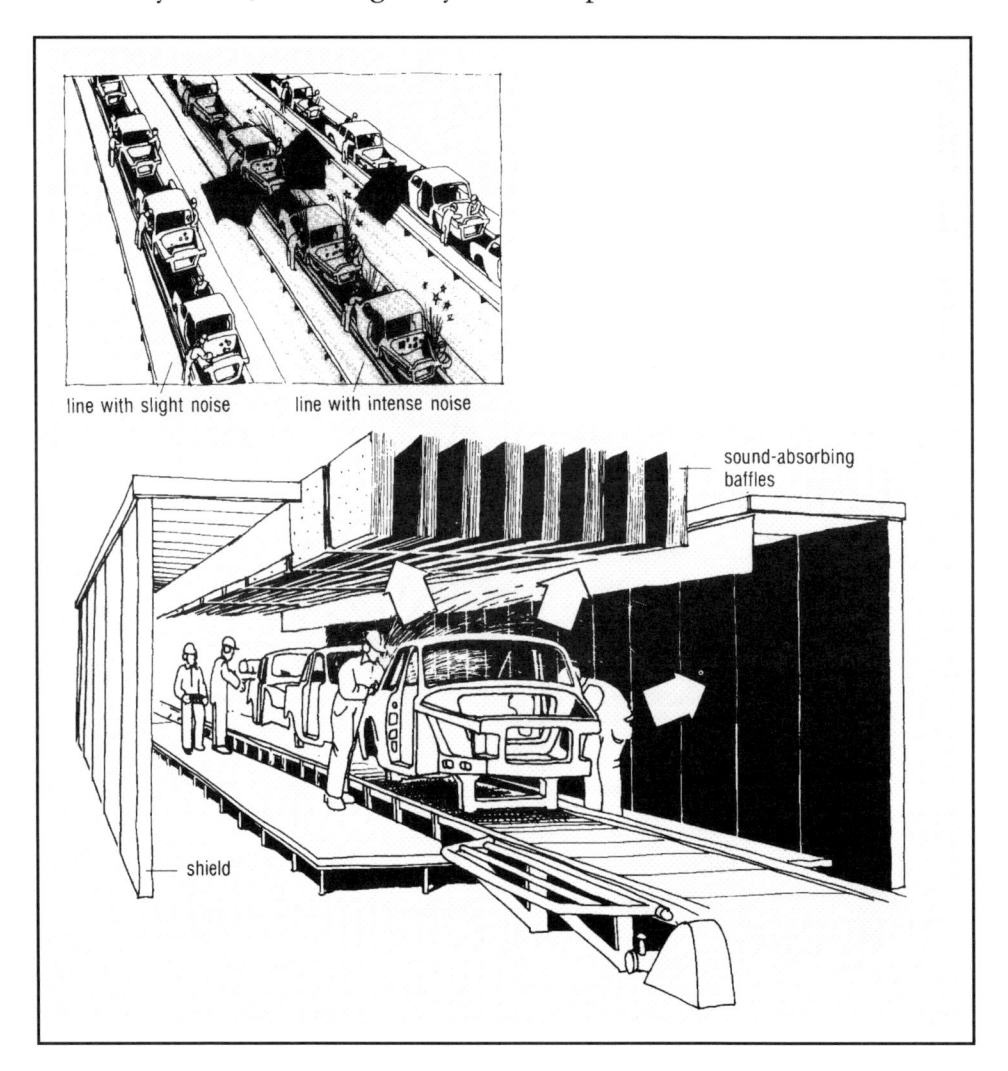

Control Measure

The other lines are protected from the grinding noise by means of shields on both sides of the line and sound-absorbing baffles suspended from the open area.

Sound Sources Should Not Be Placed Near Corners

The closer to reflecting surfaces a sound source is placed, the greater the noise it will radiate to a given distance. The worst placement is in corners near three surfaces. The best placement is away from the walls.

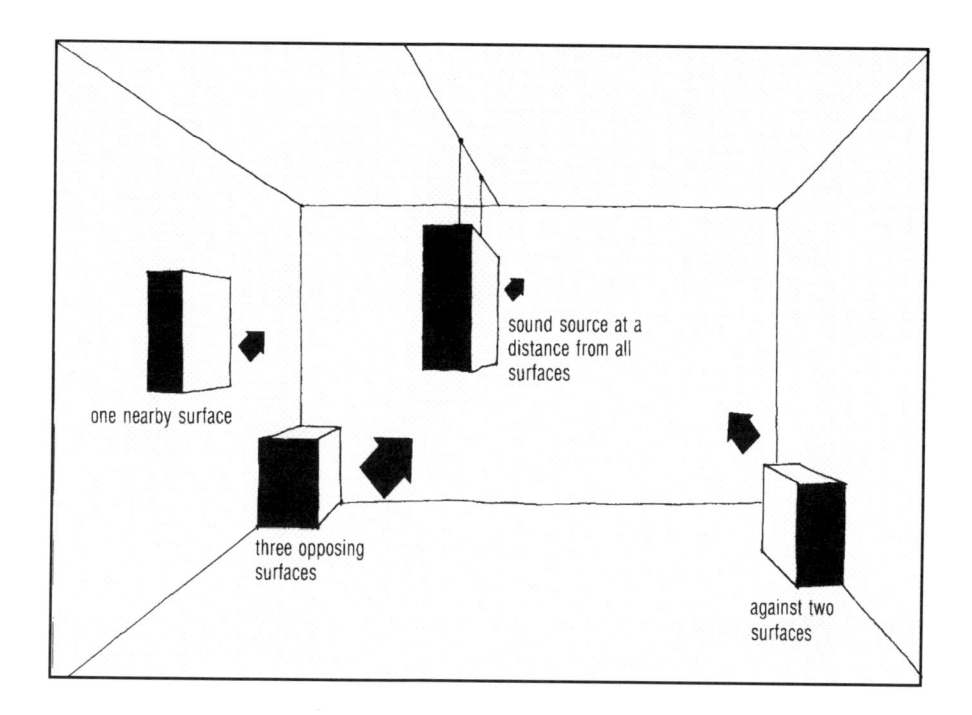

Principle

sound source at a distance from all surfaces

one nearby surface

three opposing surfaces

against two surfaces

Application of Noise Control Principles

Example

In an industrial shop, machines are placed in four rows with three aisles between them. This arrangement increases the noise from the machines in the two outermost rows.

Control Measure

The machines were placed together, two by two, away from the walls, and new aisles are set up along the walls.

Sound Movement in Ducts

All Duct Changes Reduce Sound Transmission

With all changes in a pathway, some sound energy is reflected back. In a duct this may apply to bends, branches, and changes in volume, shape or wall materials.

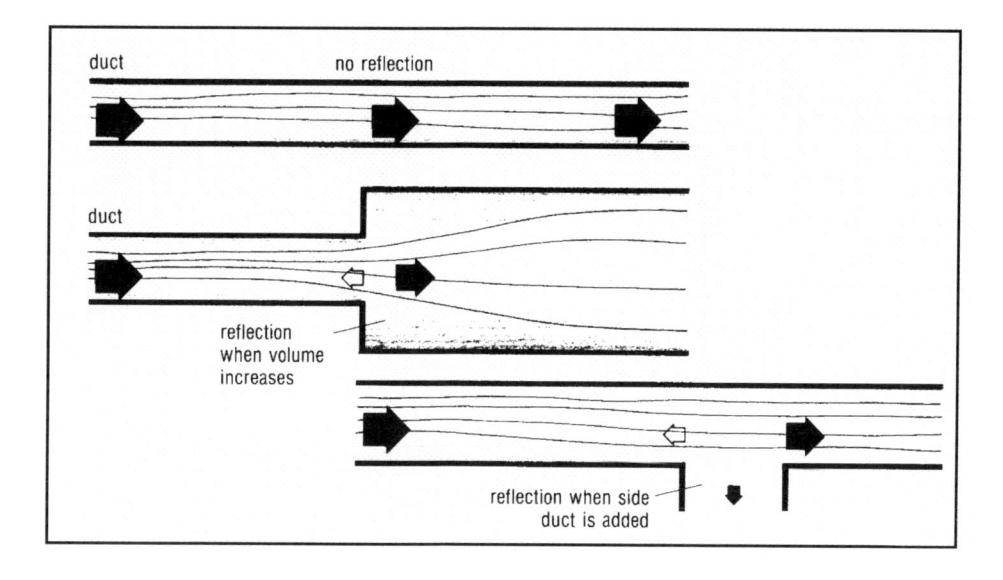

Principle

Example

An area is to be provided with mechanical ventilation. There is sufficient space for the fan, but not for the necessary muffler.

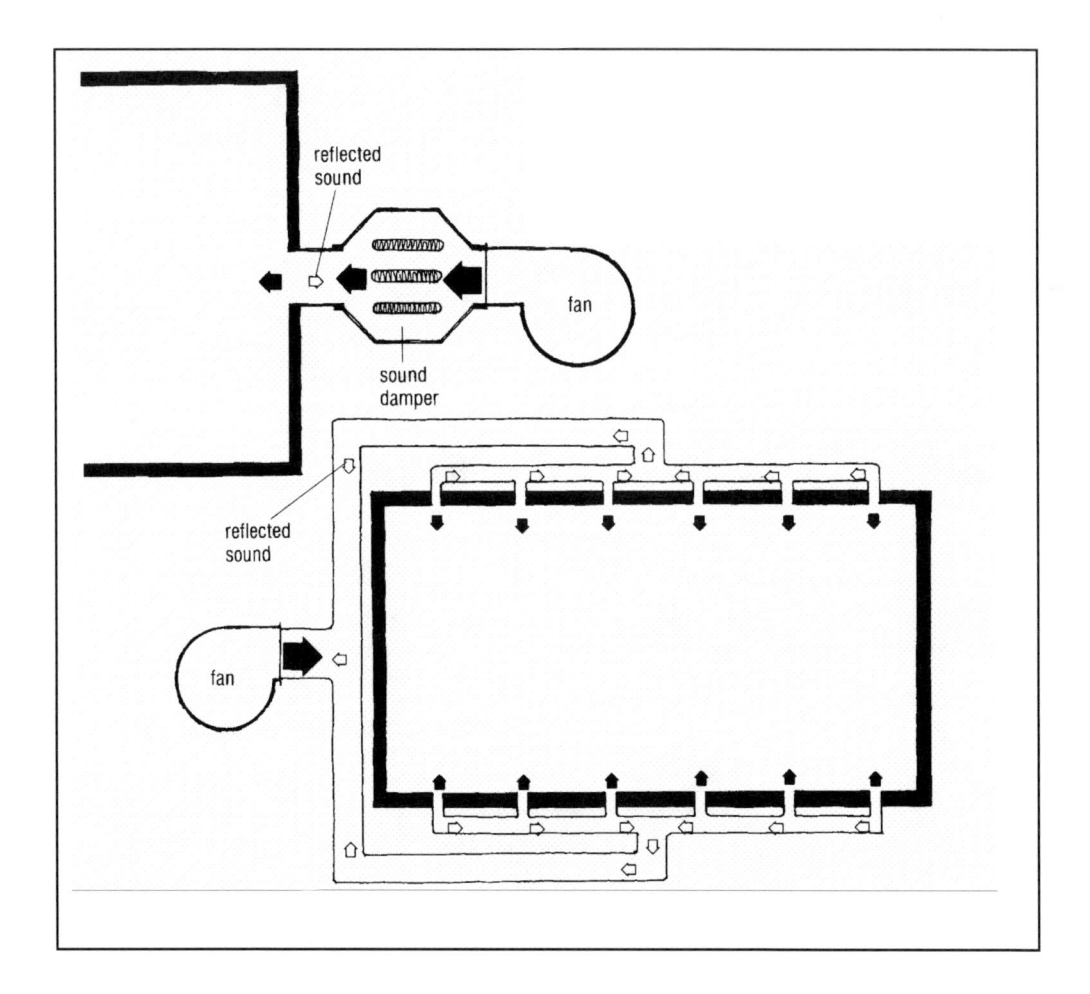

Control Measure

In place of a single inlet into the room, several smaller inlets are used. The sound reflection that takes place with all of the changes in volume and at each bend replaces the muffler.

Expansion Chambers Are Useful for Reducing Low-Frequency Noise

If a duct is provided with an expanded section or chamber, the low-frequency pressure variations in the duct are reduced. The lower the frequency which must be reduced, the greater the space required in the chamber.

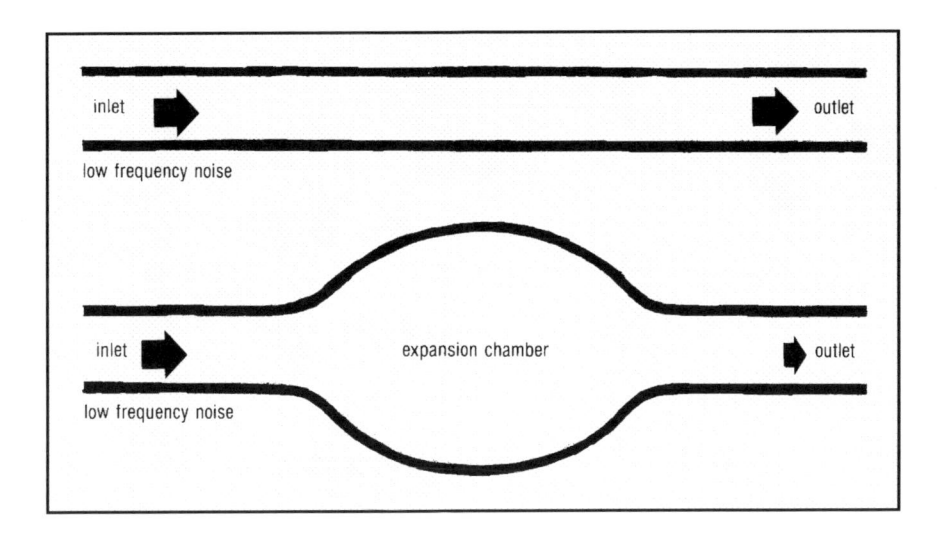

Principle

Example

Using a jacket over the tip and a tubular outlet in the jacket, the high-frequency noise given off by a jack hammer can be partially shielded. The low-frequency noise in the exhaust air is effectively reduced. The enlarged area between the barrel and jacket functions as an expansion chamber.

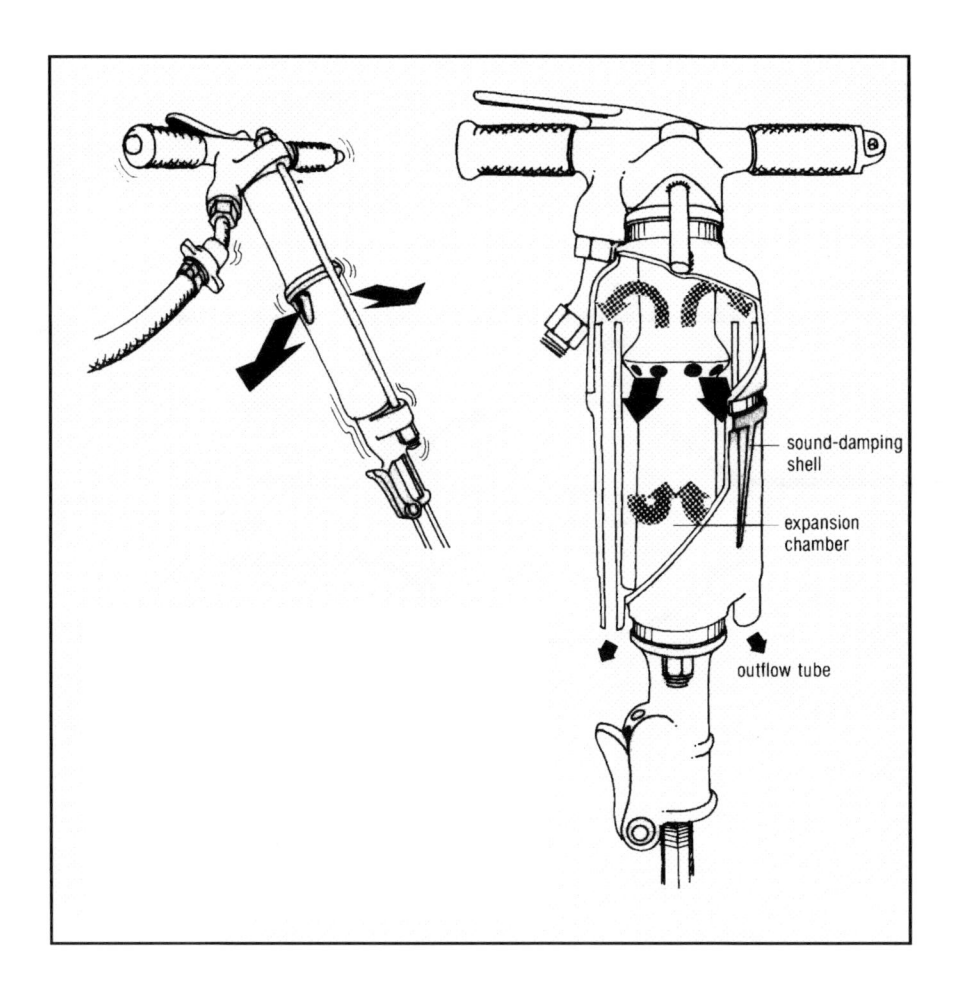

Reflection Mufflers Are Effective in Narrow Frequency Ranges

If noise is present in a limited frequency range, a reactive muffler may take up the least space. These are generally used at low frequencies. A large frequency range can be covered using several reactive chambers in series. Perforated tubes are also employed in reflection dampers.

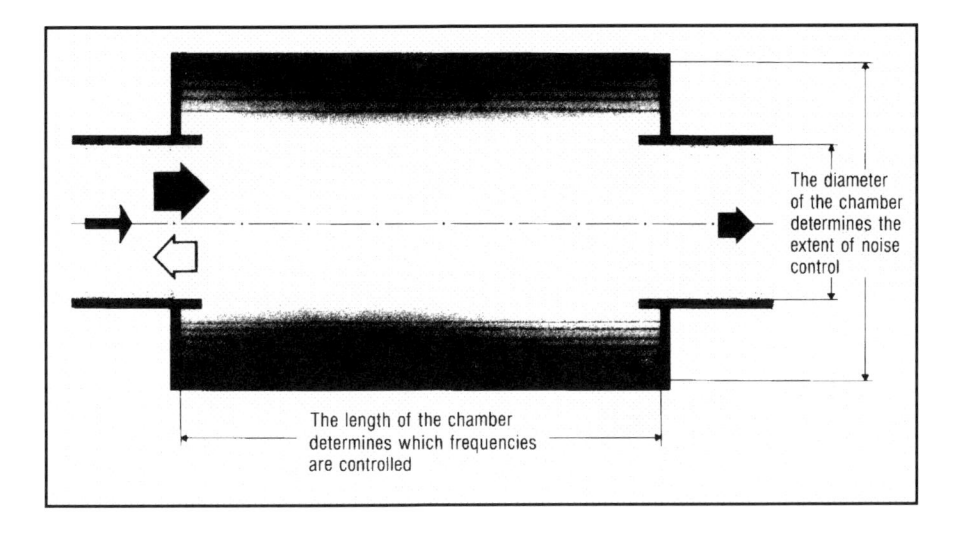

Principle

The diameter of the chamber determines the extent of noise control

The length of the chamber determines which frequencies are controlled

Example

The muffler shown here is primarily used in large piston engines.

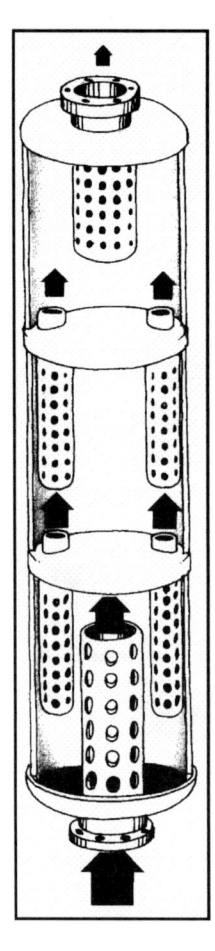

Reactive muffler with three stages.

Absorption Mufflers Are Effective Over a Broad Range of Frequencies

The simplest form of absorption muffler is a duct with sound-absorbing material on the inner walls. The thicker the material, the lower the frequency that can be reduced. For higher frequencies, the space between the absorbing walls must be made smaller. A large duct must therefore be subdivided into many smaller ones.

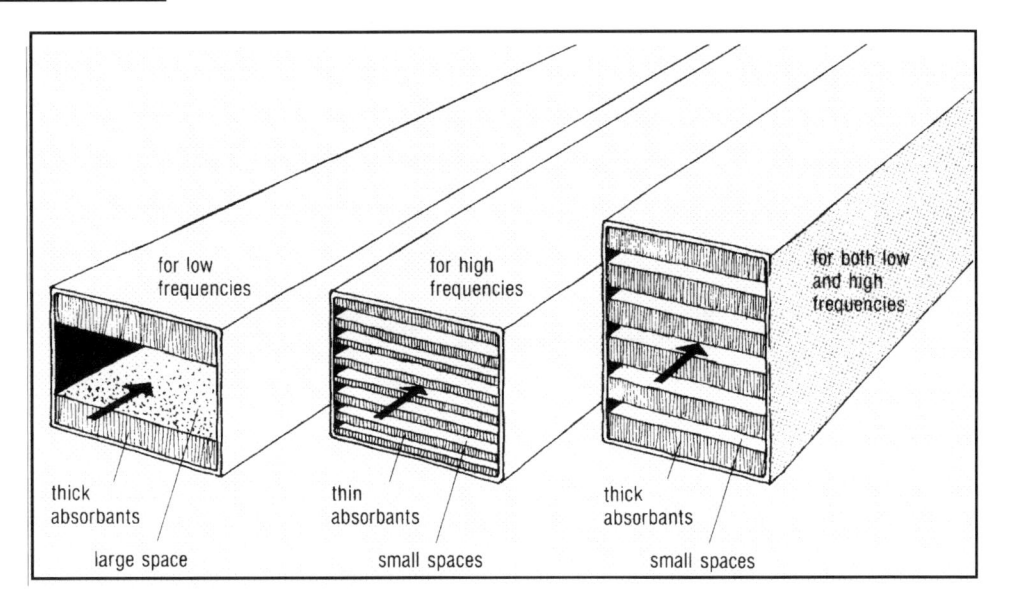

Principle

Application of Noise Control Principles

Example

If a very large frequency range is to be reduced, it is generally necessary to employ absorption mufflers with thick and thin baffles.

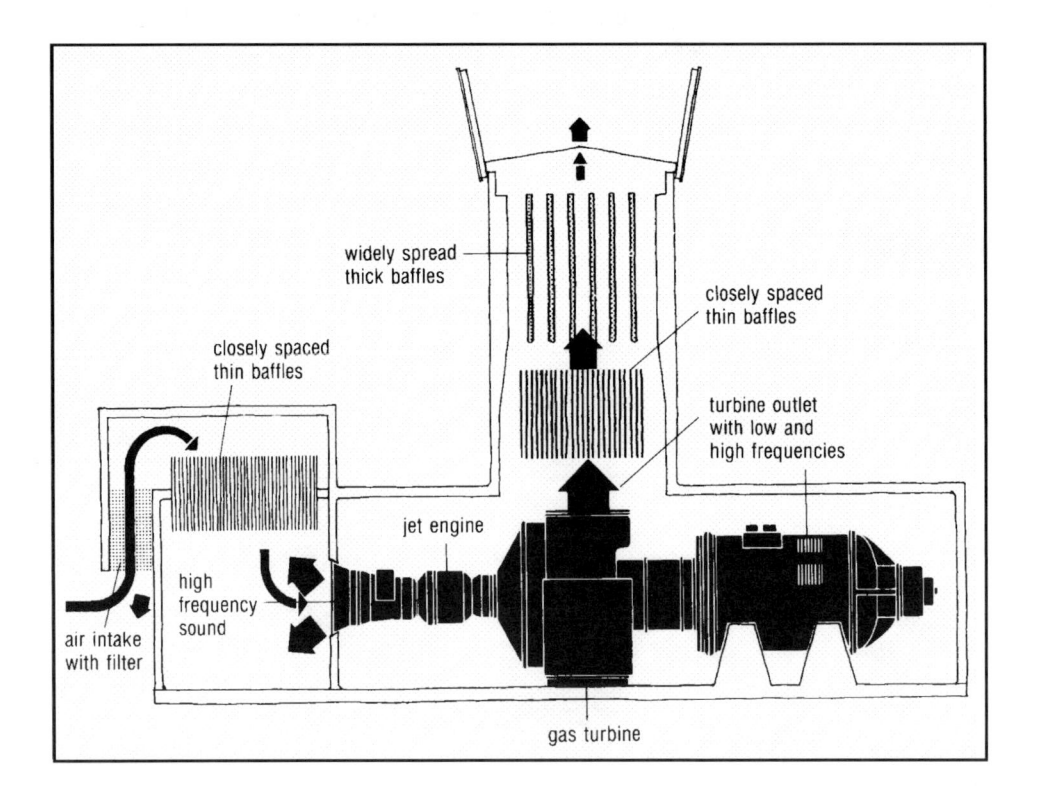

Unused Areas Can Be Absorption Chambers

The absorption chamber is a simple muffler. One section of the duct is made up of a room whose walls are covered with sound-absorbing material. When the sound is reflected against the chamber walls, sound energy is absorbed. To prevent the direct passage of high-frequency, directed sound, the inlet and the outlet should not be located opposite one another. The greater the chamber volume and the thicker the absorbent used, the lower the frequency at which the muffler is effective.

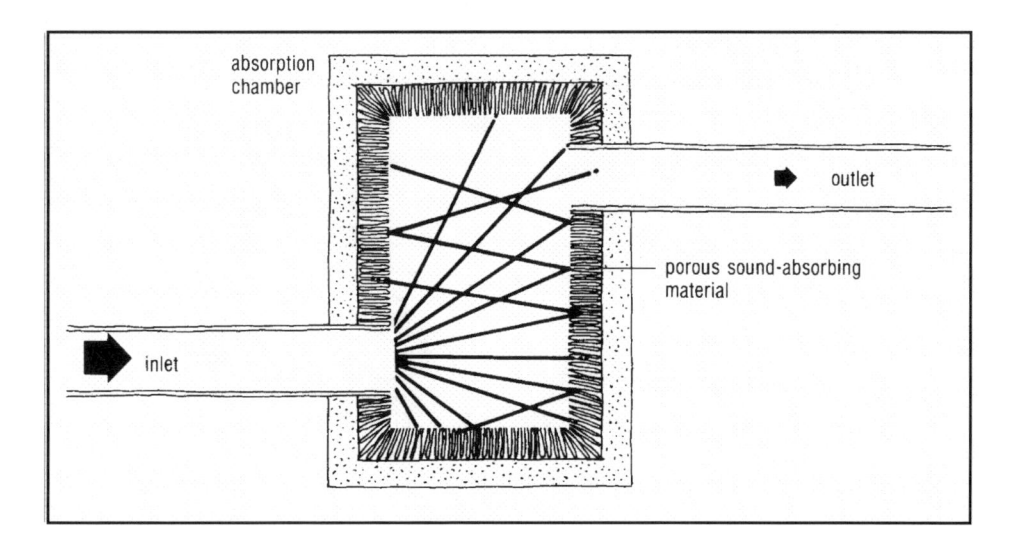

Principle

Example

The shape of the absorption chamber is of little significance. Unused rooms can be simply converted to absorption chambers.

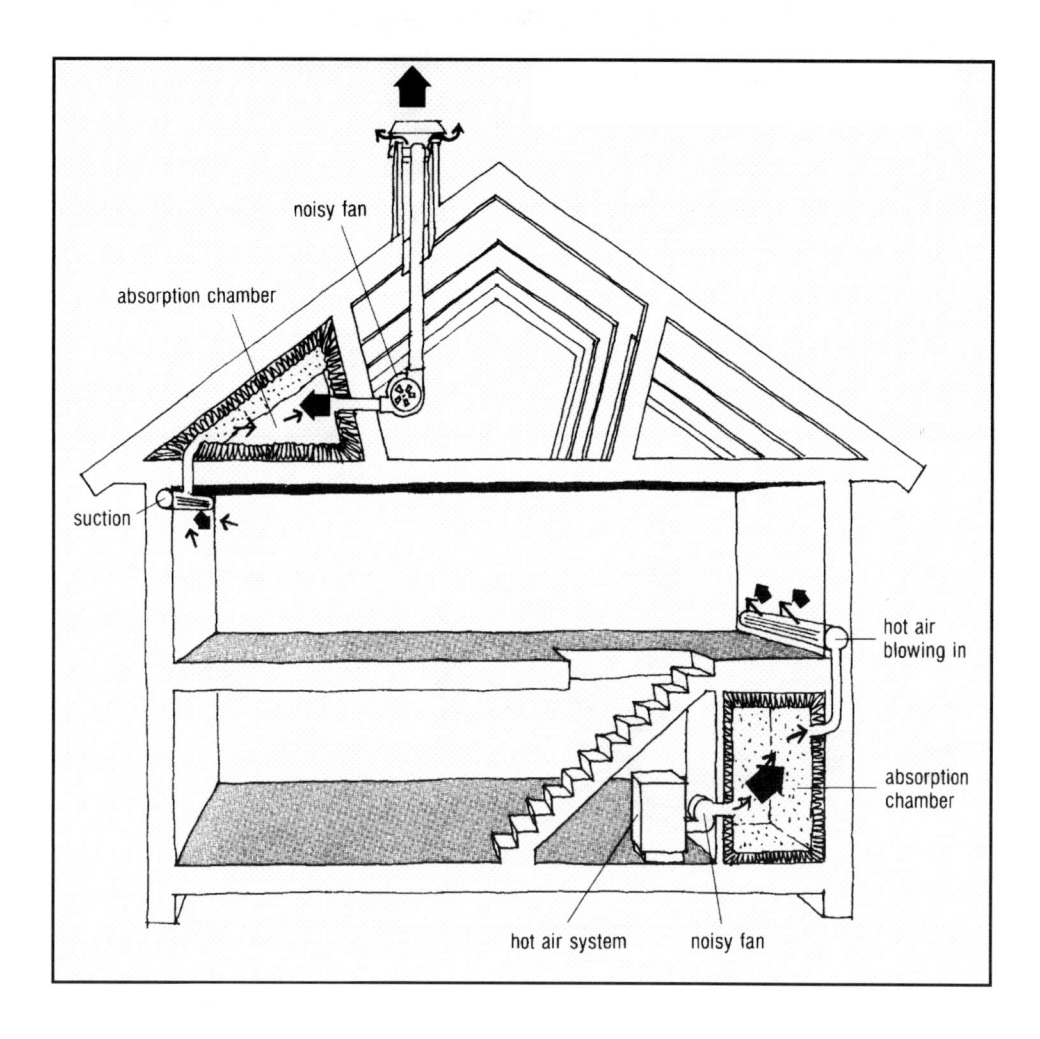

Sound From Vibrating Machines

Machines That Vibrate Should be Mounted on Heavy, Rigid Bases

Knocking on a thin door produces more sound than knocking on a thick wall. For the same reason, noise sources should be mounted on heavy or rigid bases.

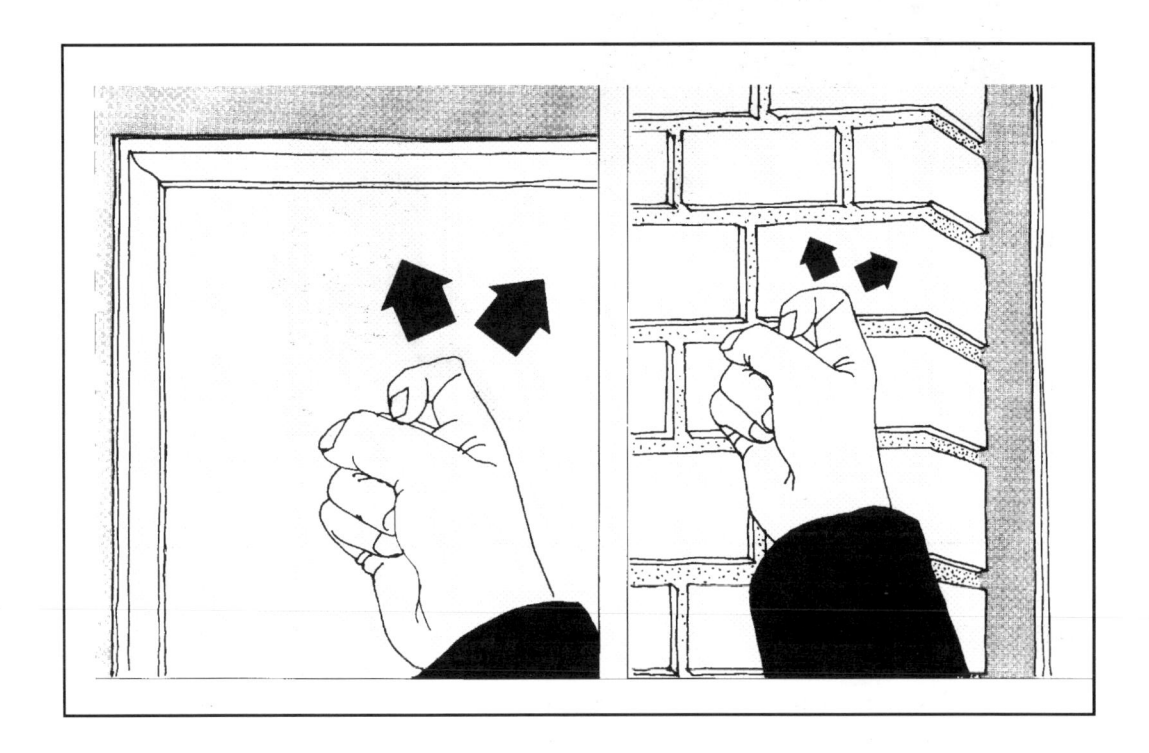

Principle

Example

A motor-driven oil pump is placed on the side wall of a hydraulic press. The vibrations are transmitted to wall plates, which convert the solid-borne sound to loud airborne sound.

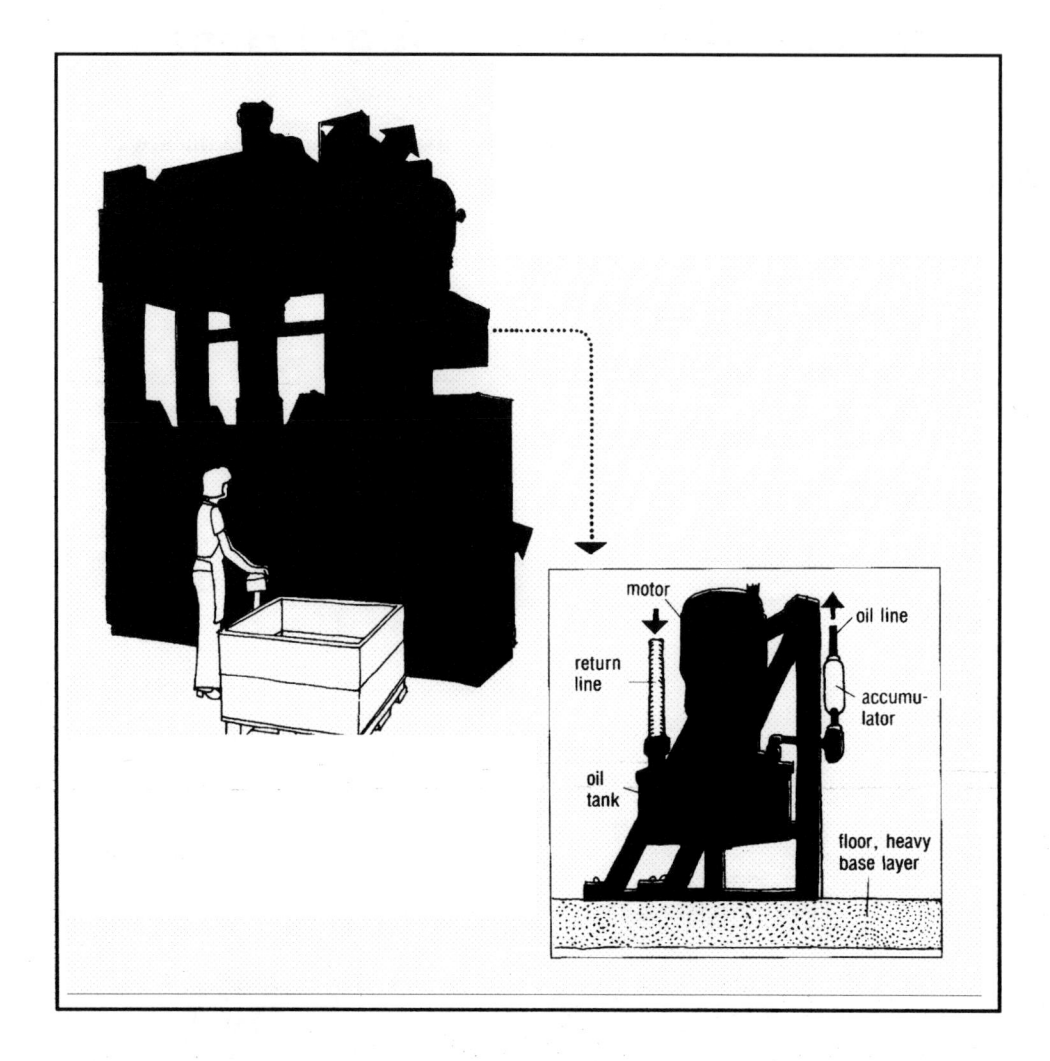

Control Measure

The oil system is removed from the press and installed in a frame on a heavy base. Sound transmission in the oil line is controlled with an accumulator.

Machines Should Be Vibration Isolated

Vibration isolation of machines can reduce the area of excessive noise as shown below. Either the machine or the working area can be isolated.

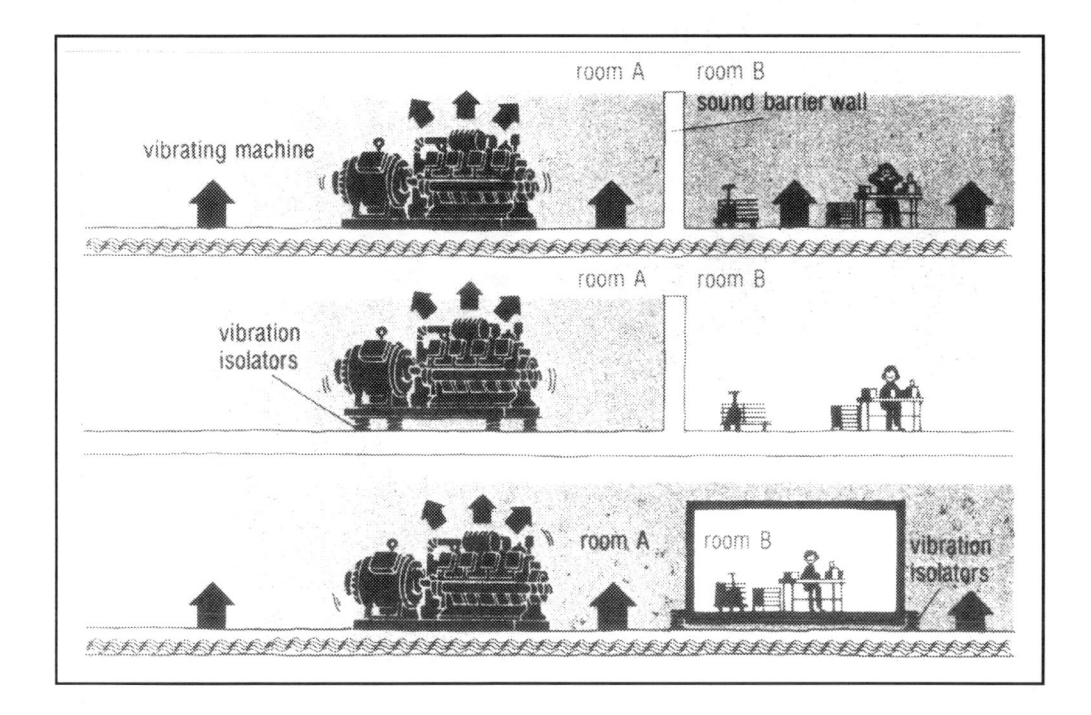

Principle

Application of Noise Control Principles

Example

Vibration isolators are made of various materials and in various shapes.

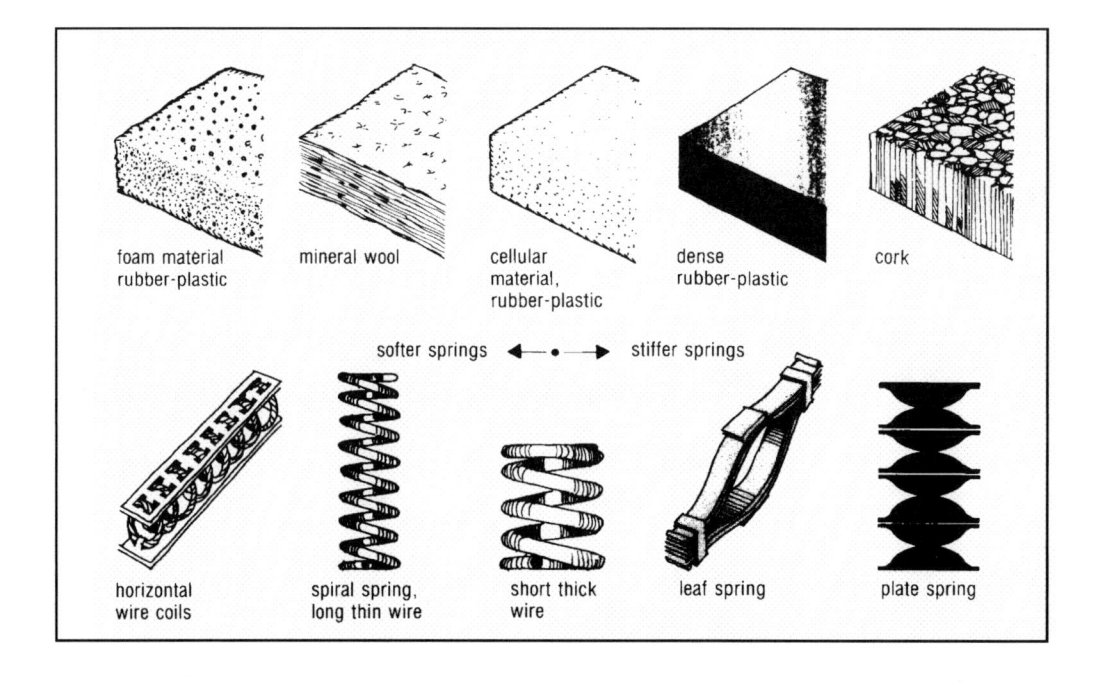

| foam material rubber-plastic | mineral wool | cellular material, rubber-plastic | dense rubber-plastic | cork |

softer springs ◄—•—► stiffer springs

| horizontal wire coils | spiral spring, long thin wire | short thick wire | leaf spring | plate spring |

Improperly Selected Springs Can Increase Vibrations

A machine placed on springs has a so-called "fundamental frequency." Vibrations at or close to the fundamental frequency are greatly intensified. The machine may even break away from its fastenings. Vibrations with lower frequency than the fundamental frequency are not blocked. If the base is very heavy or very rigid, the fundamental frequency is determined entirely by the machine and base weights together with the rigidity of the spring. The lighter the machine and the more rigid the spring, the higher the fundamental frequency. This reinforcement of vibrations can be avoided by using springs with good internal damping.

Principle

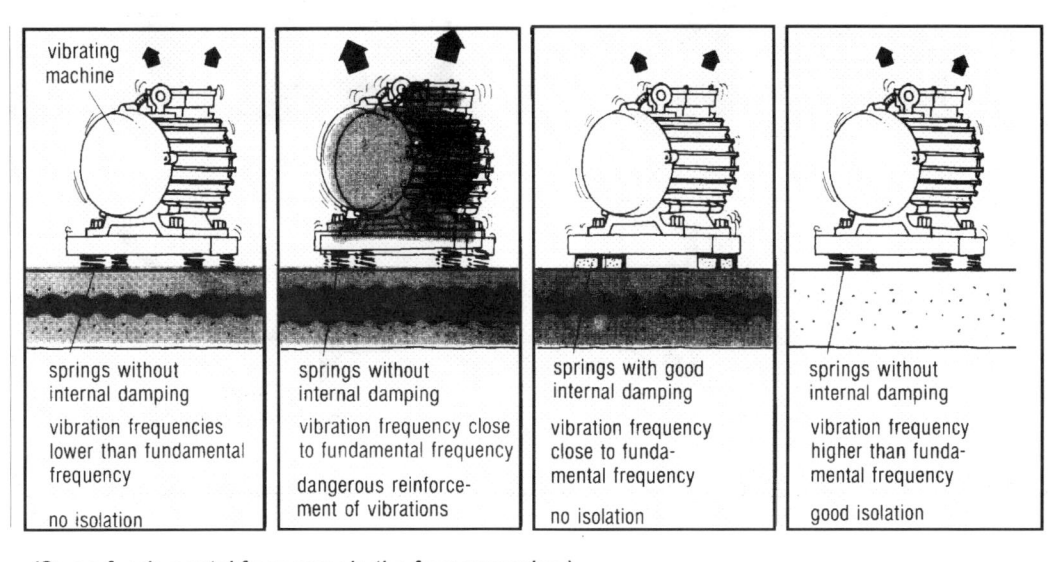

springs without internal damping	springs without internal damping	springs with good internal damping	springs without internal damping
vibration frequencies lower than fundamental frequency	vibration frequency close to fundamental frequency	vibration frequency close to fundamental frequency	vibration frequency higher than fundamental frequency
no isolation	dangerous reinforcement of vibrations	no isolation	good isolation

(Same fundamental frequency in the four examples.)

Example

Two fans are used in the same building. Both are vibration isolated with steel springs which have very poor internal damping. The isolation functions well for both fans during constant operations, but one of the fans is started and stopped frequently. When this happens, the vibration frequency corresponds for a short time with the fundamental frequency, which produces serious disturbance.

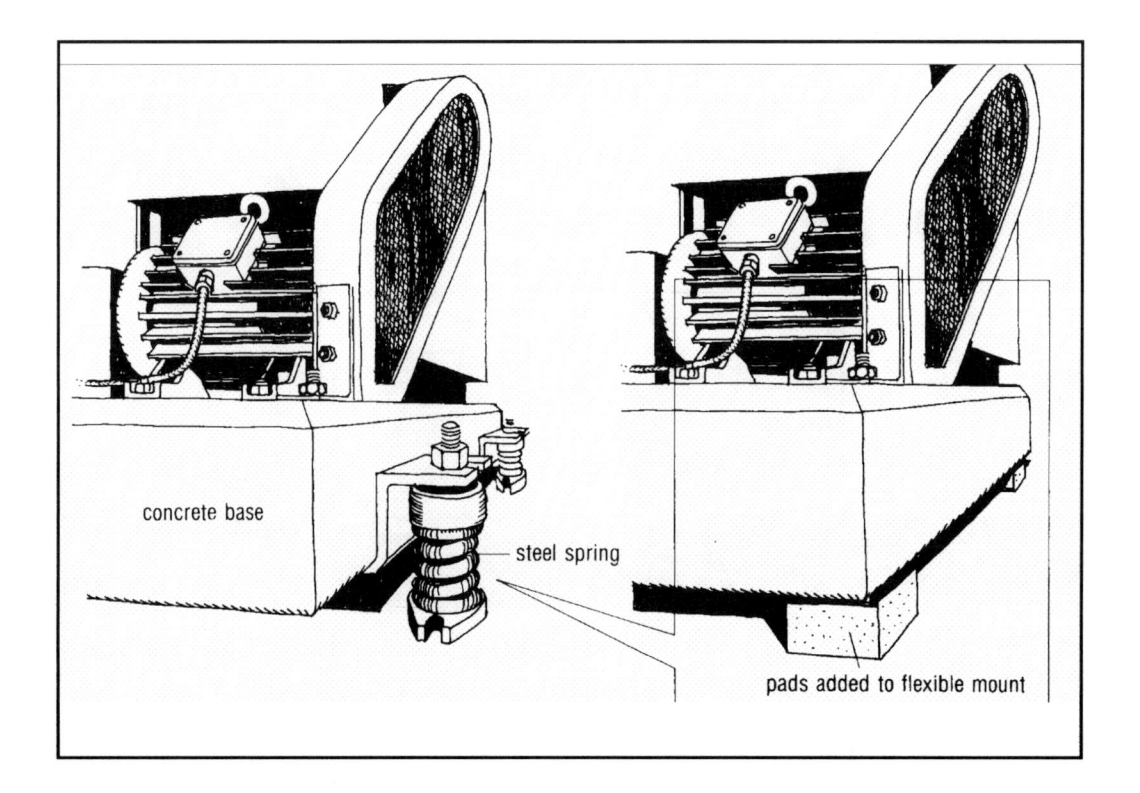

concrete base — steel spring — pads added to flexible mount

Control Measure

On the fan with irregular operation, steel dampers are installed with pads which have good internal damping. The isolation is somewhat less, but the disturbance from starting and stopping disappears.

Isolating Machines with Low Natural Frequency May Require a Rigid Floor

A machine and mount with low natural frequency are difficult to vibration isolate unless the floor is very rigid. As shown below, an extra heavy (stiff) or pile-reinforced floor might be necessary.

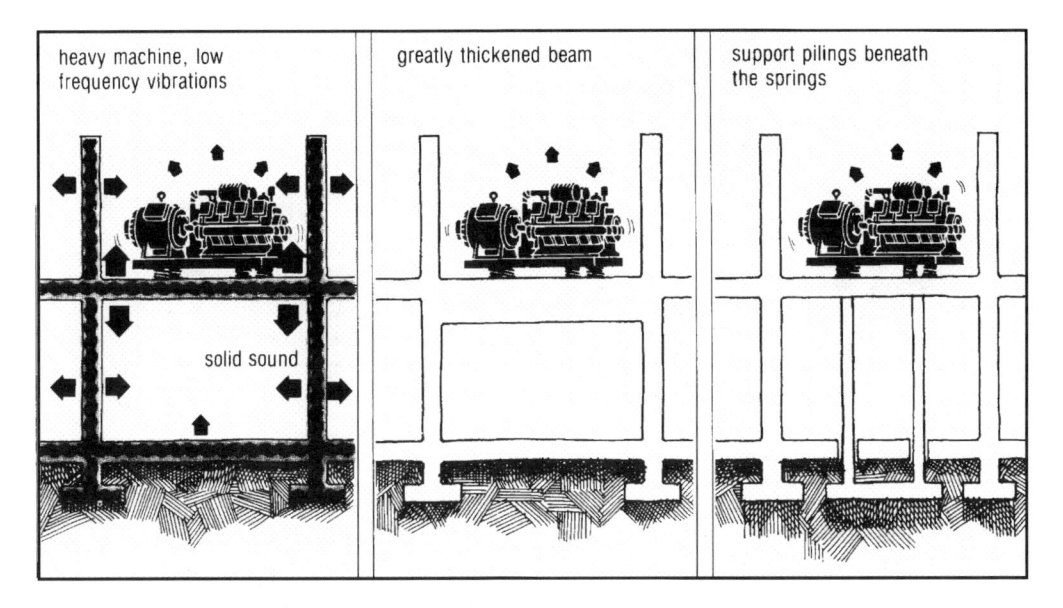

Principle

Example

A company is planning a building where a need for freedom from vibration and noise is great. It should also be possible to remove and interchange the machines.

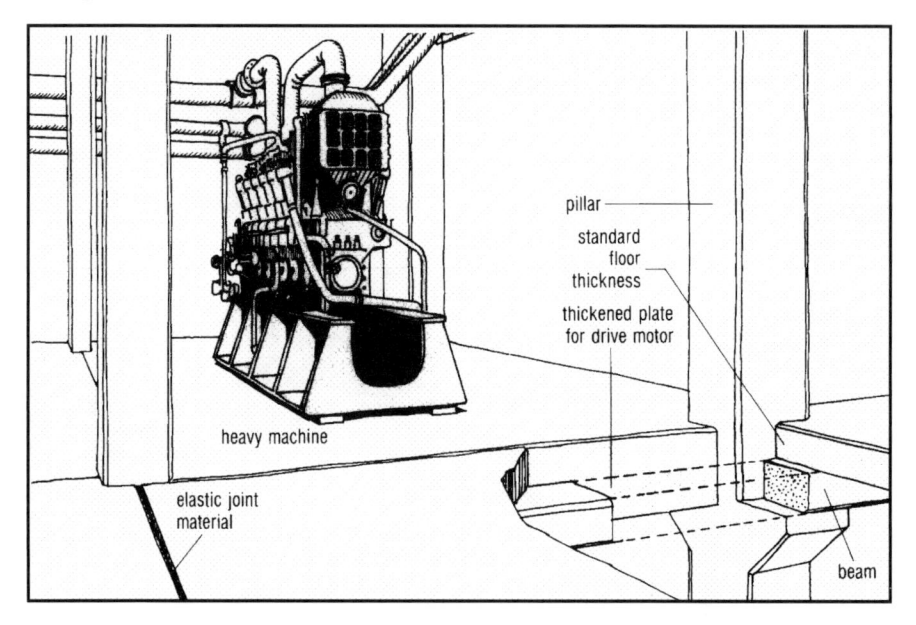

Control Measure

The building is constructed with large concrete plates on a pillar and beam system. The concrete plates which are expected to carry heavy machines are provided with strong reinforcements. If heavy machines are added later, the normal concrete plate is removed and replaced with a thicker one.

A Separate Base Layer Provides the Best Solid-Borne Sound Barrier

A good way to isolate very heavy machines with low natural-frequency vibration is to place them on a concrete base plate which rests directly on the ground. Even more effective protection is achieved if the base plate is separated from the remainder of the building by means of a joint. If the ground has a clay layer, it may be necessary to place pilings beneath the plate.

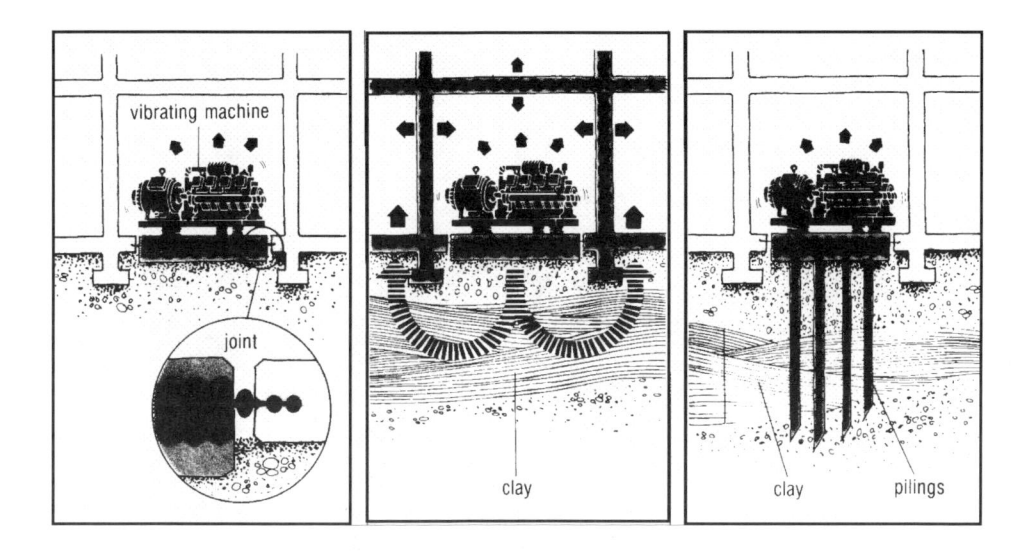

Principle

Example

Drive motors with gears and differentials connected to a paper-making machine cause both loud air noise and vibrations in the machines. They require only occasional maintenance which can generally be performed with the machines turned off. Therefore, the machines can be permitted to make large amounts of noise if the noise is prevented from entering the rest of the factory.

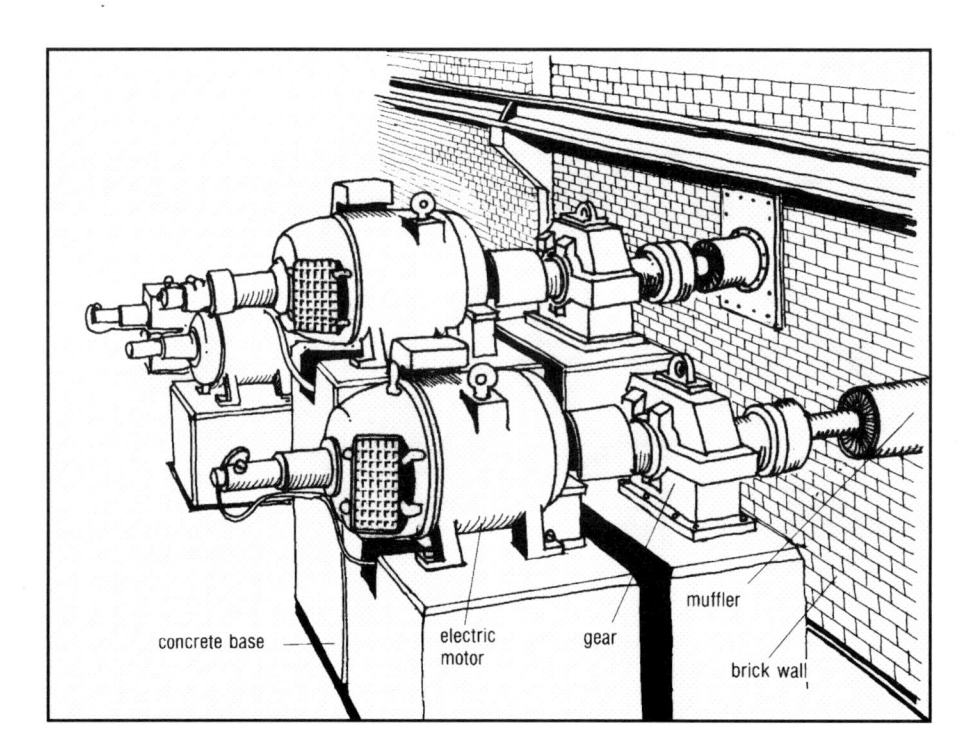

Control Measure

The engine room has its own thick base plate which is in good contact with the solid ground. The large base plate is also vibration isolated with corrugated rubber mates. Sound is prevented from entering other rooms by means of a brick wall. Holes in the wall for the axles to pass through are sealed with mufflers.

Sound Through Solid Connections Can Be Blocked

Vibration isolation of a machine may be ineffective if sound is transferred through connections for oil, electricity, water, etc. These connections must be made very flexible. The machine movements will be reduced if a heavy base is selected, and more rigid springs can be used.

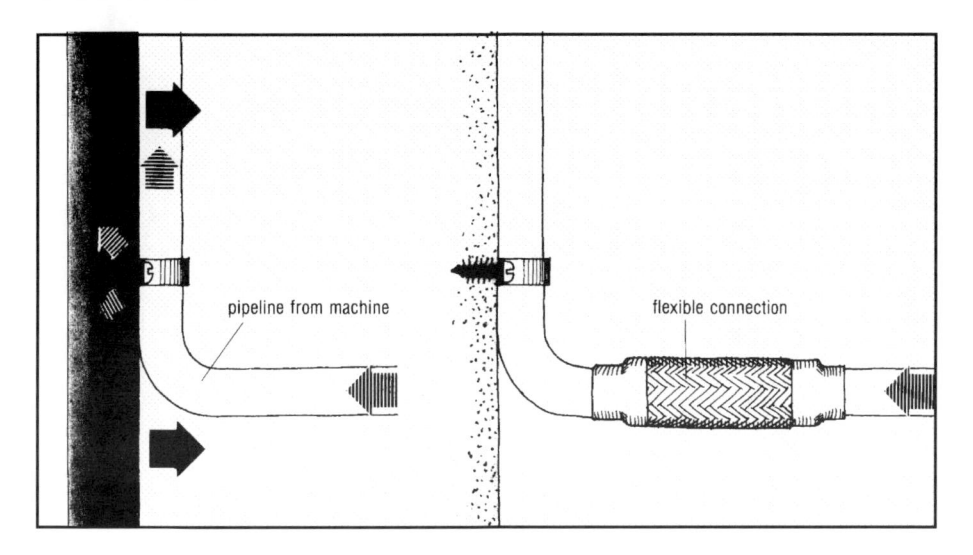

Principle

pipeline from machine

flexible connection

Example

Cooling systems may be serious sources of noise as a result of intense pressure shocks in the liquid from compressors.

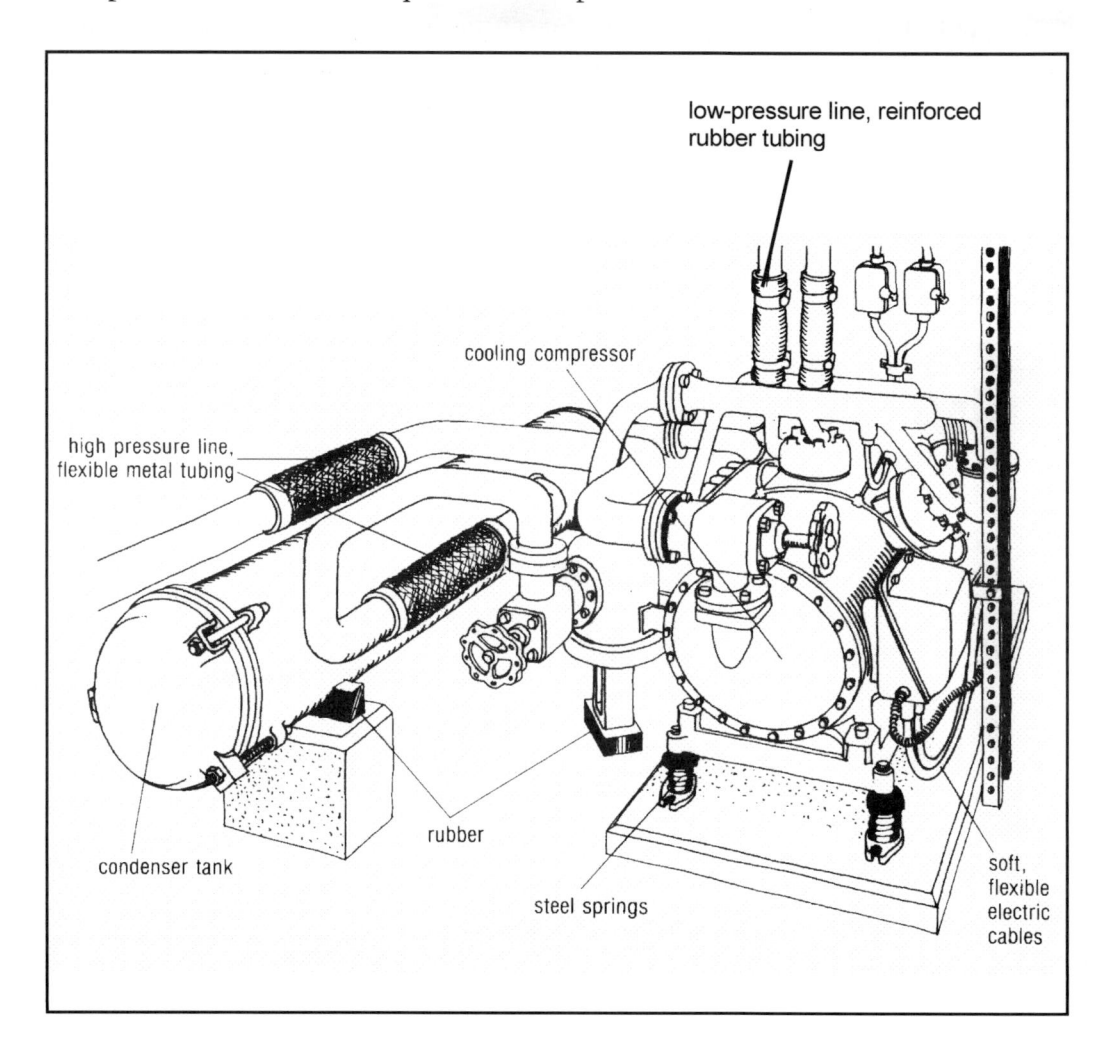

Control Measure

Compressors may be vibration isolated with steel springs. In addition, flexible connections should be used for all inlet and discharge pipes.

Sound Reduction in Enclosure Walls

The TL of a Single Wall is Estimated from Its Surface Weight

"Transmission Loss" (TL) indicates a wall's ability to absorb sound-producing vibrations. TL is expressed in decibels (dB). The TL of a homogeneous single layer wall can be estimated by its surface weight, that is, kilograms per square meter (kg/m^2) or pounds per square inch (lbs/in^2).

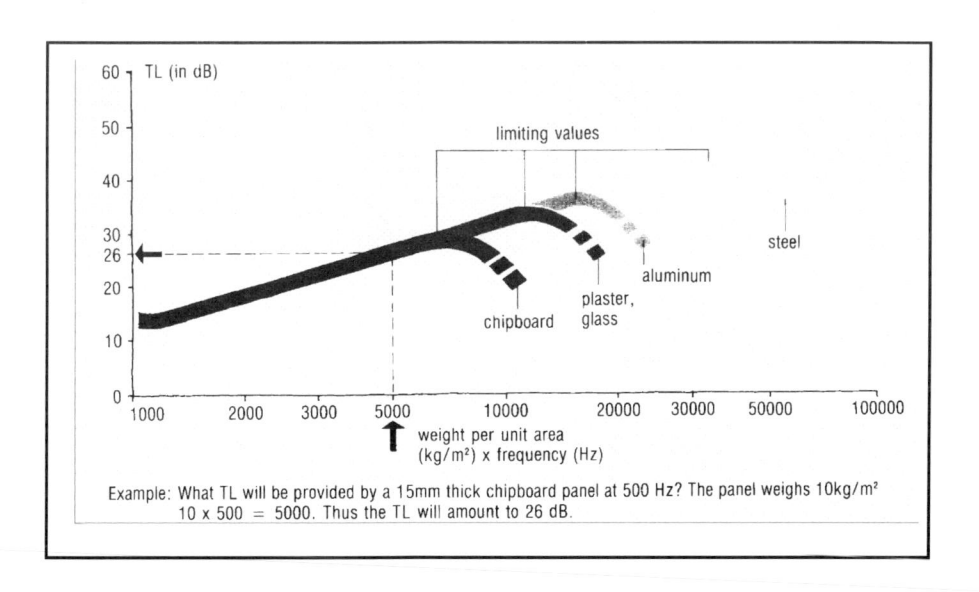

Principle

Example: What TL will be provided by a 15mm thick chipboard panel at 500 Hz? The panel weighs 10kg/m²
10 x 500 = 5000. Thus the TL will amount to 26 dB.

Example

Behind one end wall in a long factory room are a number of machines with an intense noise level peak around 1000 Hz. The end of the room is isolated with a wall of 25 mm chipboard and 6 mm glass. The isolation is ineffective since the chipboard has its coincidence valley at 1000 Hz.

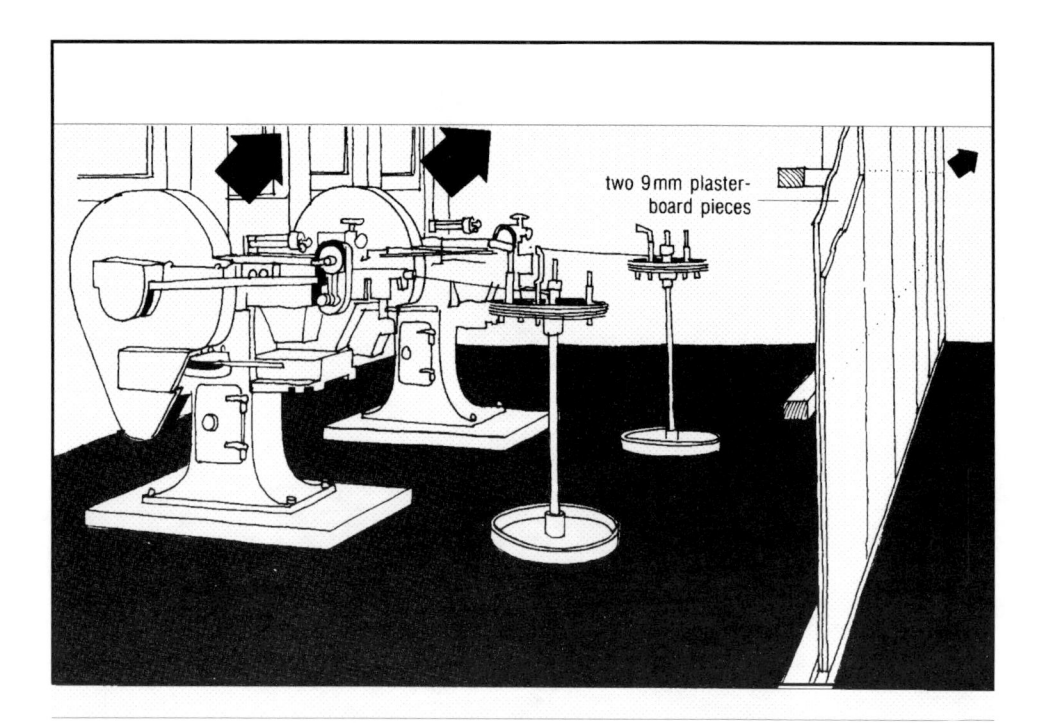

Control Measure

The chipboard is replaced by two layers of 9 mm plasterboard. The isolation is improved by about 10 dB. The plasterboard weights about the same as one 25 mm chipboard, but is less than one-fourth as rigid. The coincidence valley is located at 2500 Hz. (Absorptive material could also be added to the machine side of the wall.)

Rigidity and Weight Are Both Important in Thick Walls

In most single walls, the coincidence valley is close to 100 Hz for a thickness of about 20 cm. At higher frequencies, both increased weight and increased rigidity produce greater TL. A cast concrete wall has greater rigidity than a brick wall, and therefore provides greater TL if the two wall weights are equal.

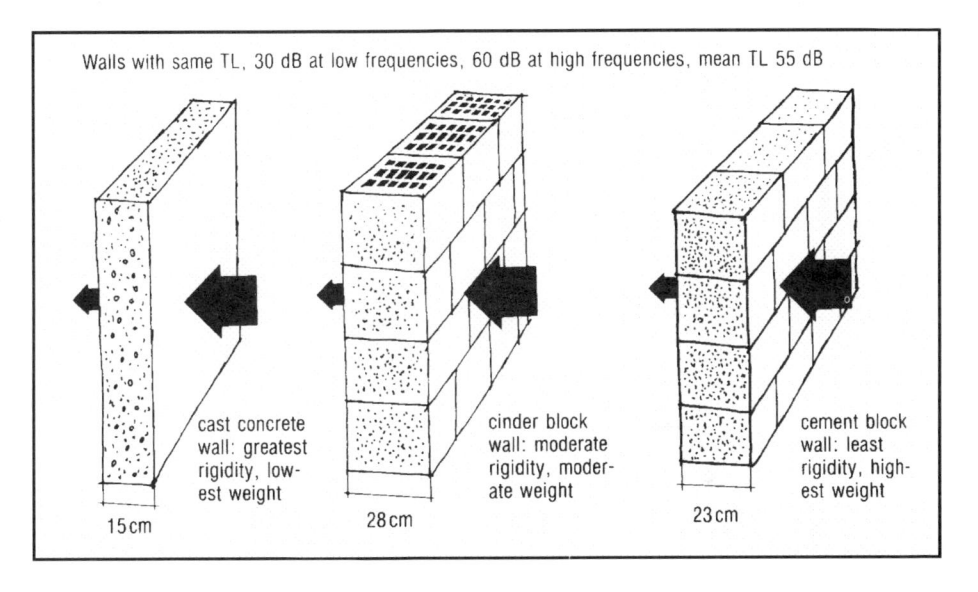

Principle

Walls with same TL, 30 dB at low frequencies, 60 dB at high frequencies, mean TL 55 dB

- cast concrete wall: greatest rigidity, lowest weight — 15 cm
- cinder block wall: moderate rigidity, moderate weight — 28 cm
- cement block wall: least rigidity, highest weight — 23 cm

Example

Machines in a large open area in an industrial building create a noise hazard.

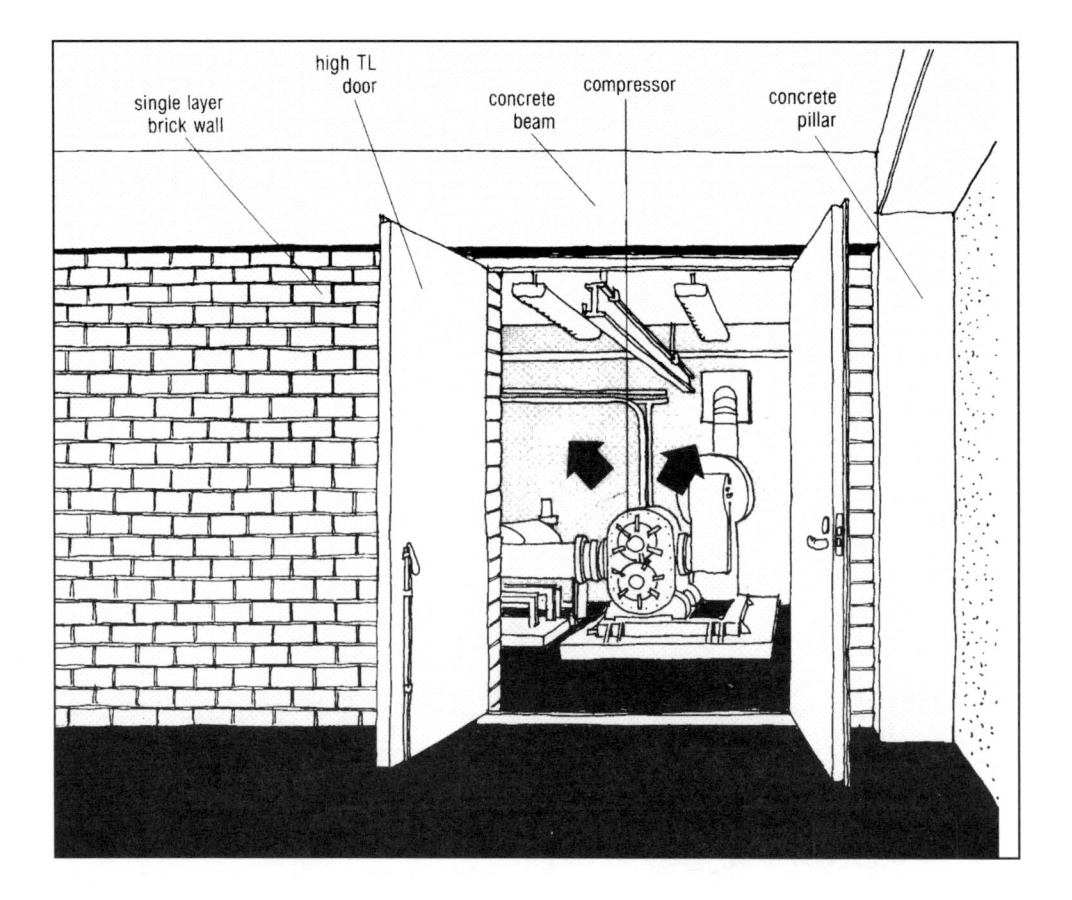

Control Measure

The area containing the machines is surrounded by a brick wall.

Light Double Walls Provide Good TL

Two light walls separated by an air gap provide good TL, increasing with the distance between them to about 15 cm. With sound-absorbing material in between, the TL further increases as the distance between increases. Double walls may provide the same TL as single walls that are five to ten times as heavy.

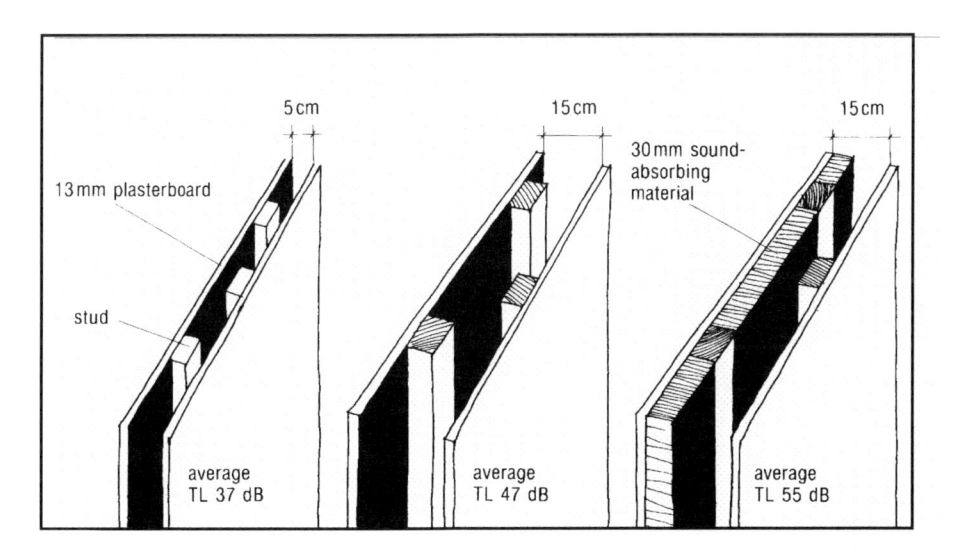

Principle

Example

Extremely noisy machines disturb workers in adjacent work areas separated from the noisy equipment by a thin wall.

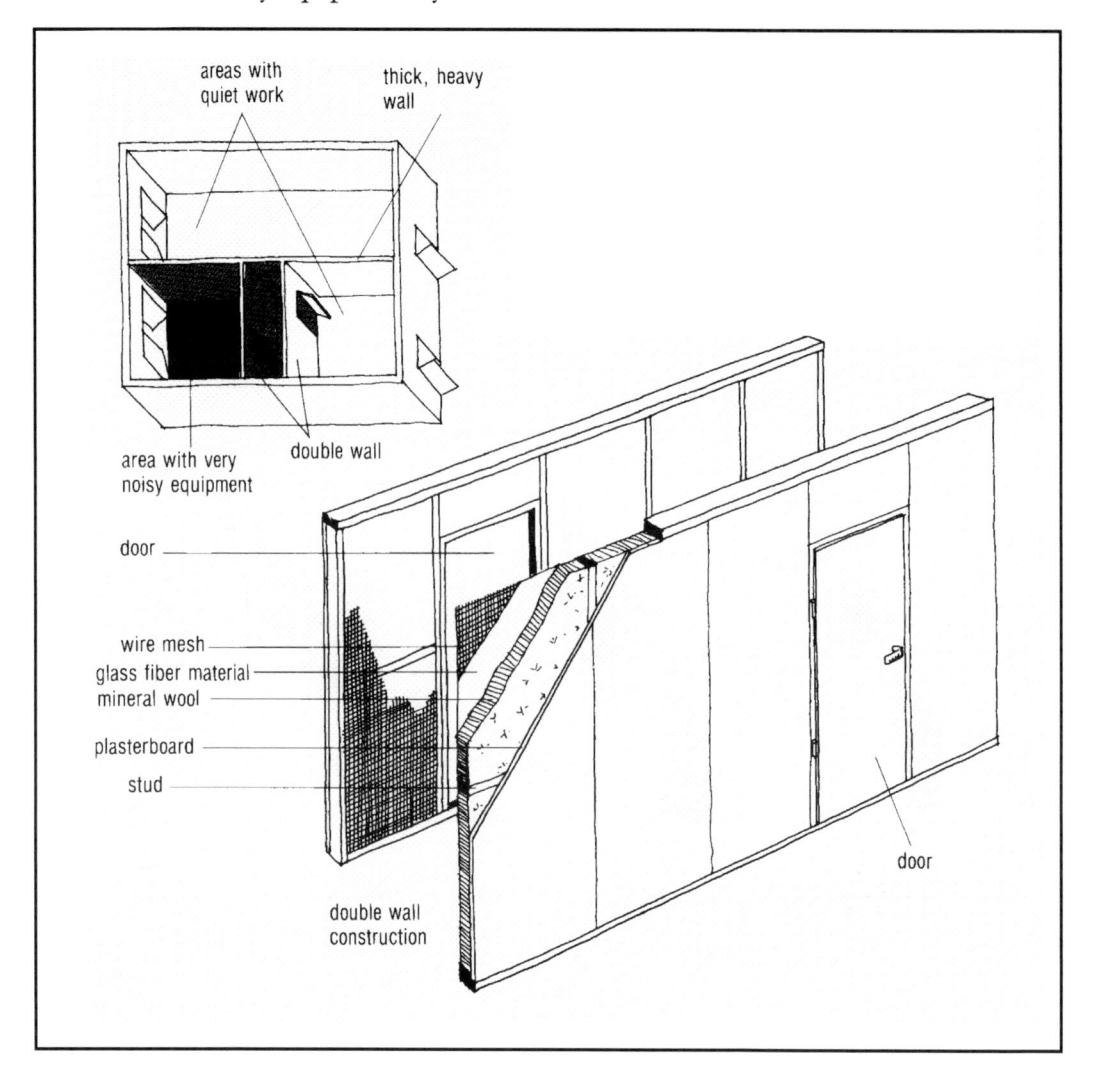

Control Measure

The machines are brought together and placed in one end of the factory. This is separated off with a light double wall that provides 60 dB of insulation. A passageway is created with doors to the two quiet areas. The insulation between the areas is at least 35 dB even if one of the doors is open.

Double Walls Should Have Few Connections

A double wall provides the best TL if each layer is connected to heavy walls or if there are open joints on both ends. If the layers are fastened to shared studs, the TL is greatly reduced if the studs are close together. The thicker the layers, the farther apart the studs must be in order to avoid substantial reduction of TL

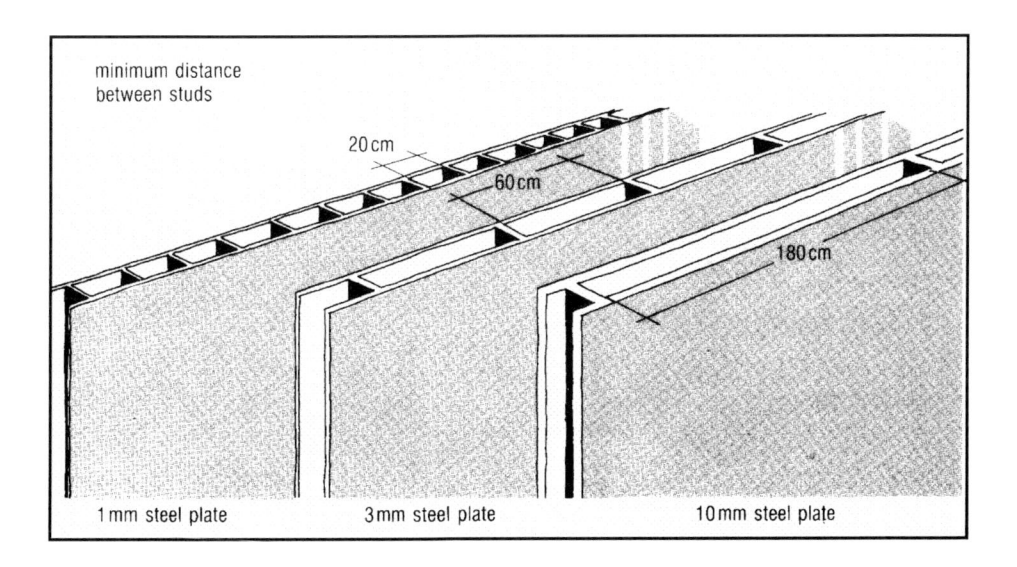

Principle

Application of Noise Control Principles

Example

The control room for a machine in a paper mill is noisy, and telephone conversation is practically impossible.

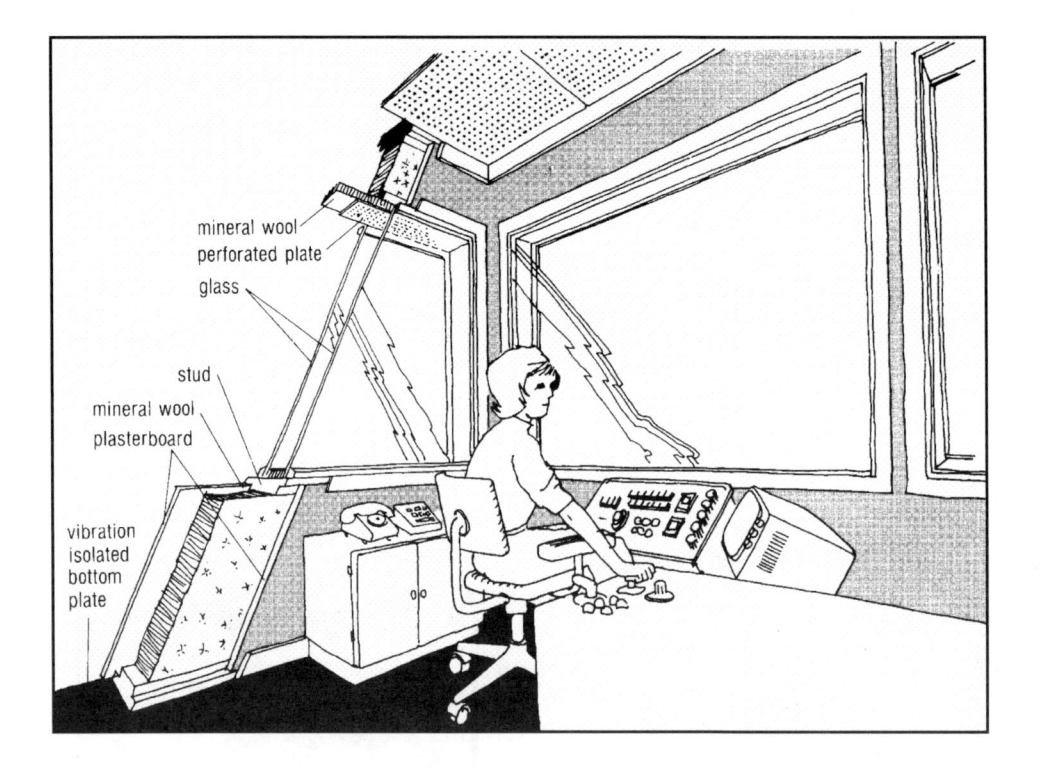

Control Measure

A well-insulated room is created with thin panels on common studs. The floor plate of the room is isolated from vibration from the floor of the factory.

Noise Control Measures

The following is a survey of noise control methods which have been applied with good results in various types of workplaces. Many noise sources produce airborne sound and sound from vibrating surfaces at the same time, so in may cases several noise control measures must be applied.

Changes in Machinery and Equipment

The machines or machine parts to be controlled must be identified. Methods of maintenance and servicing must be taken into accountin noise control design. Attempts should be made to:

- Prevent or reduce impact between machine parts
- Gently reduce speeds between forward and reverse movements
- Replace metal parts with quieter plastic parts
- Enclose especially noisy machine parts.

Designers should be encouraged to:

- Select power transmission which permits the quietest speed regulation, e.g., rotation-speed-controlled electric motors
- Isolate vibration-related noise sources within machines
- Provide proper TL and seals for doors of machines
- Provide machines with effective cooling flanges which reduce the need for air jet cooling.

Noise levels of existing equipment can often be controlled as effectively as new equipment without complicated procedures. Common control measures include:

- Providing mufflers for the air outlets of pneumatic valves
- Changing the pump type in hydraulic systems
- Changing to quieter types of fans or placing mufflers in the ducts of ventilation systems
- Providing mufflers for electric motors
- Providing mufflers or intakes for air compressors.

In a new plant, it is sometimes possible to make more extensive changes during the design process, such as:

Noise Control Measures

- Specifying quiet electric motors and transmissions
- Selecting hydraulic systems that have remote oil tanks and accumulators at pump discharges, and designing pipelines for low flow speeds (maximum 5 m/sec.).
- Designing ventilation ducts with fan inlet mufflers and other mufflers to prevent noise transfer in the duct between noisy and quiet rooms.

Example

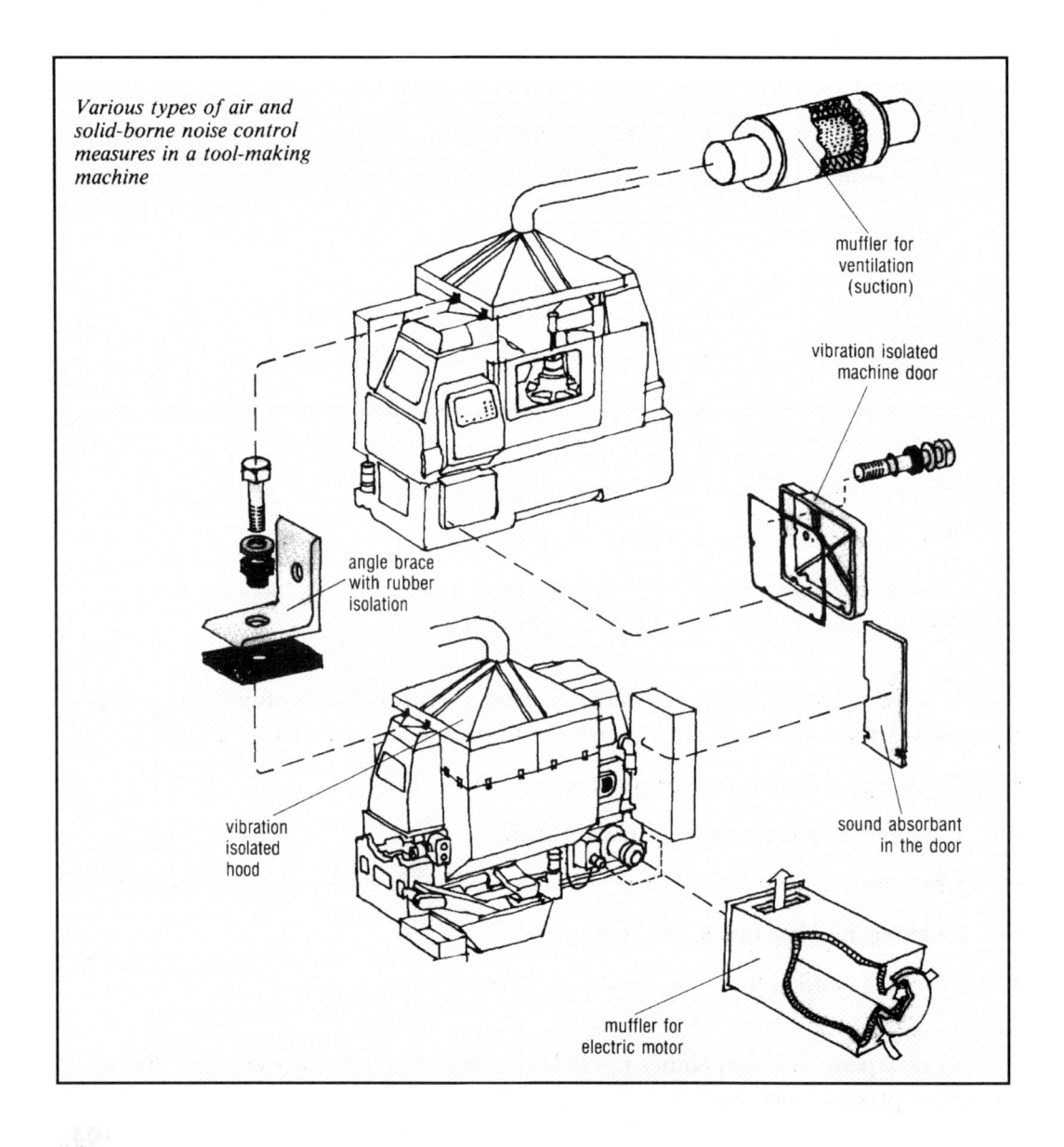

Various types of air and solid-borne noise control measures in a tool-making machine

- muffler for ventilation (suction)
- vibration isolated machine door
- angle brace with rubber isolation
- sound absorbant in the door
- vibration isolated hood
- muffler for electric motor

Materials Handling

Existing workplaces may be changed to prevent impact and collision during manual and mechanical materials handling.

- Reduce the dropping height of items being collected in bins and boxes
- Increase the rigidity of containers receiving impact from goods, or damp them with damping materials.
- Use soft rubber of plastic to receive hard impacts

If new conveying equipment is being purchased, consideration should be given to creating a system for quiet materials handling. The following may be considered:

- Selecting belt conveyors, which generally are quieter than roller conveyors
- Regulating the conveyors or other transportation systems so that their speed is adjusted to the required amount of material. In this way, it is possible to avoid some noise produced by vibrations and colliding objects.

Example

Plates dropping from a great height off of a roller belt onto a stacking platform produce loud noise.

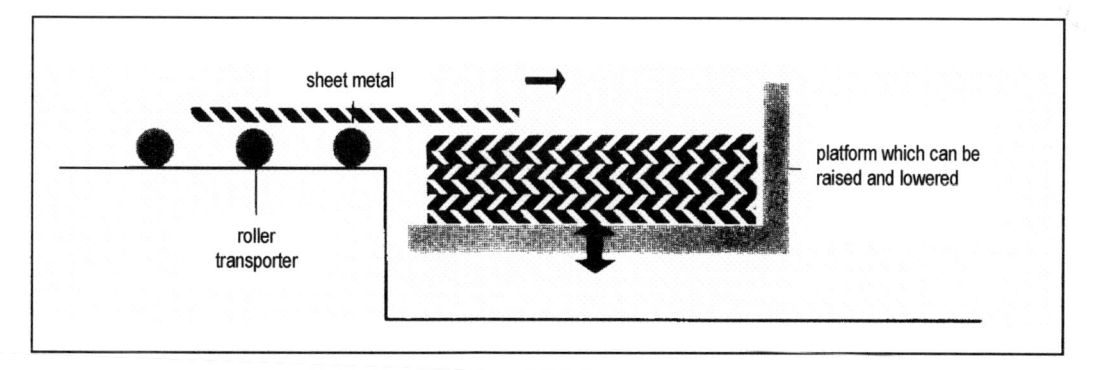

Control Measure

By using a board whose height can be raised and lowered, the drop can be reduced and the noise decreased

Enclosures of Machines

If it is not possible to prevent noise, it may be necessary to enclose the machines.

- Use a dense material, such as sheet metal or plasterboard, on the outside.
- Use a sound absorbent material on the inside. A single hood of this type can reduce the sound level by 15-20 dB (A)
- Install mufflers on cooling air openings during enclosures of electric motors, etc.
- Install easily opened doors as required for machine adjustment and service.

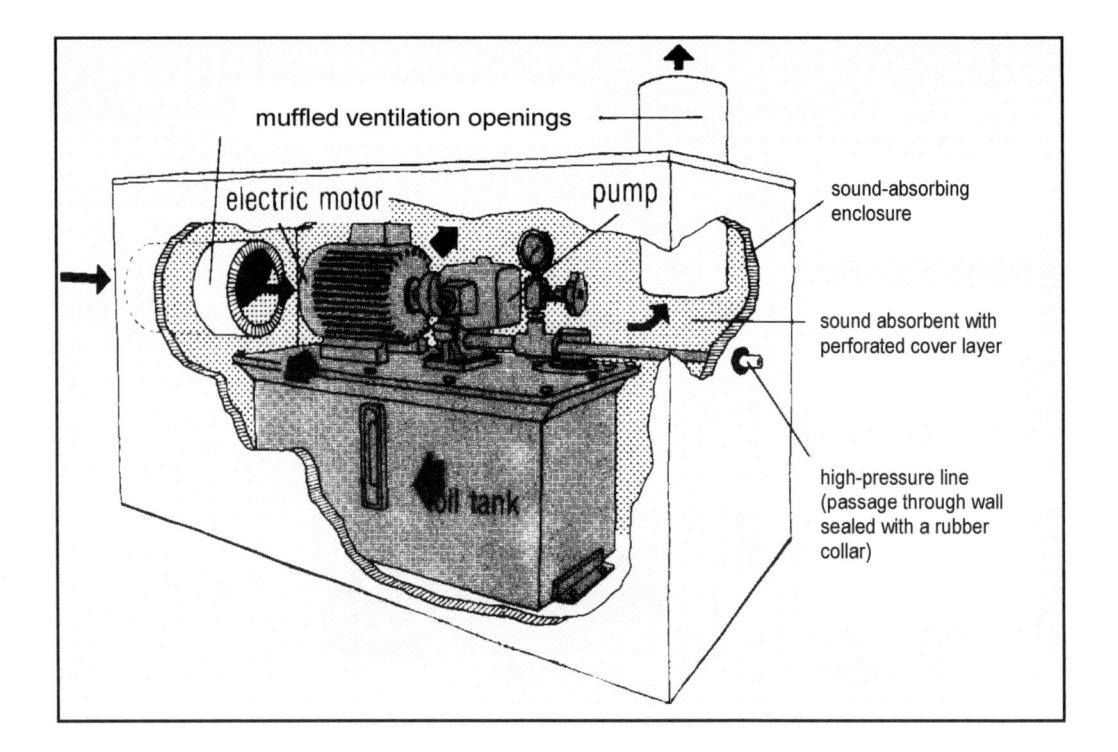

Enclosure of a hydraulic system requires muffled ventilation openings. Electric motors release both sound and heat, as do the pump and the oil tank.

Control of Noise from Vibrating Surfaces

Vibration in machines often results from slippage or loosened bolts. In such cases, the disturbance can be reduced by repair or replacement.

- Isolate the floor from machine vibrations

- Place large and heavy machines which will not be vibration isolated on separate bases. They may be put on a separate piece of ground without contact with the remainder of the building

- Provide vibration isolation of machine surfaces to reduce sound emission Fasten plates to the machine face by flexible means in order to reduce the vibrations of the surfaces. Plates with special damping design may be used.

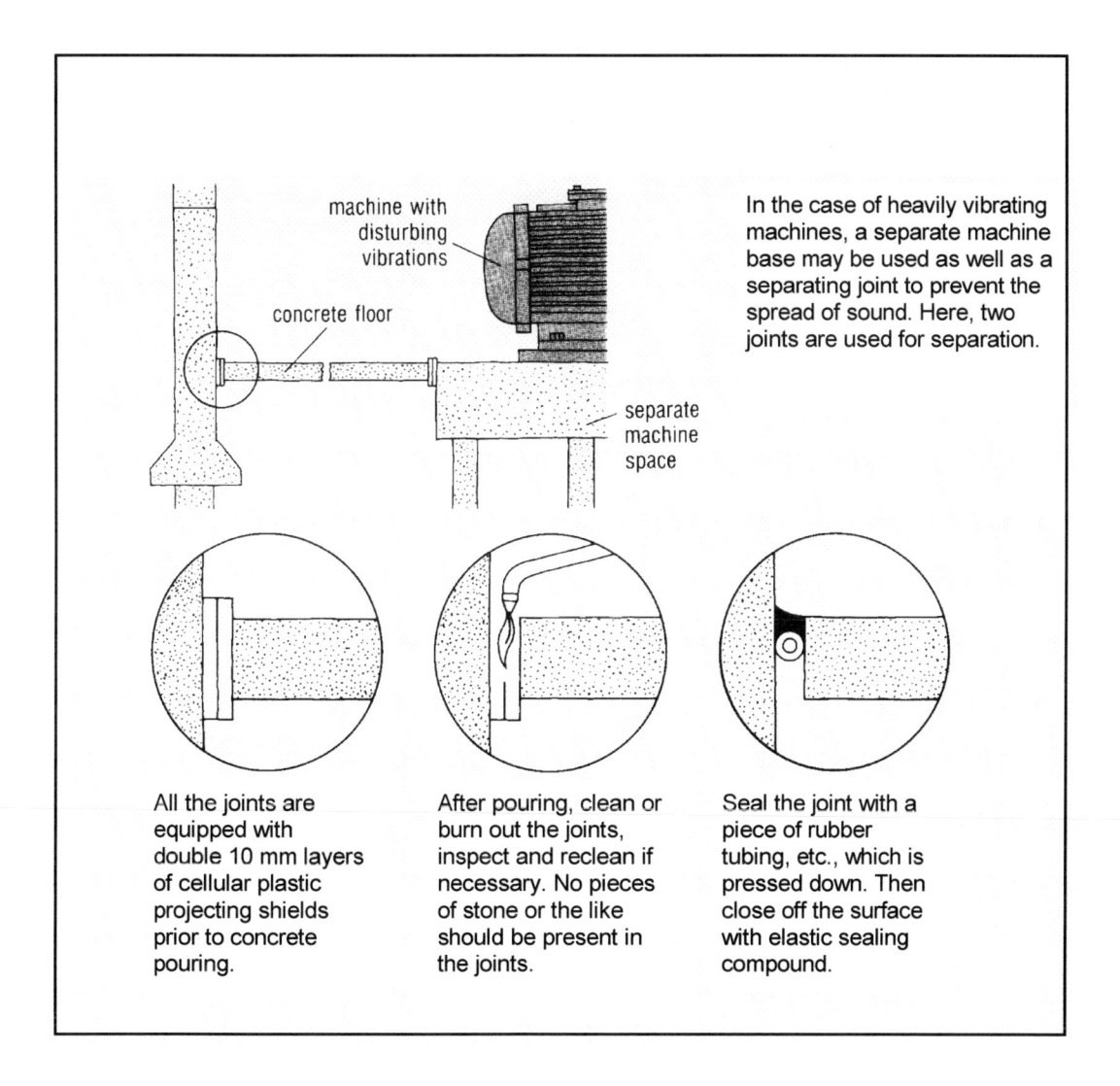

machine with disturbing vibrations

concrete floor

separate machine space

In the case of heavily vibrating machines, a separate machine base may be used as well as a separating joint to prevent the spread of sound. Here, two joints are used for separation.

All the joints are equipped with double 10 mm layers of cellular plastic projecting shields prior to concrete pouring.

After pouring, clean or burn out the joints, inspect and reclean if necessary. No pieces of stone or the like should be present in the joints.

Seal the joint with a piece of rubber tubing, etc., which is pressed down. Then close off the surface with elastic sealing compound.

Damping with Absorbents

In a workplace with hard materials on the ceiling, walls, and floor, almost all the sound which strikes the surface is reflected. The sound level goes down at first as you move away from the machine, but after a certain point it remains practically unchanged. A better sound environment can be obtained by coating the ceilings and walls with effective sound-absorbing material.

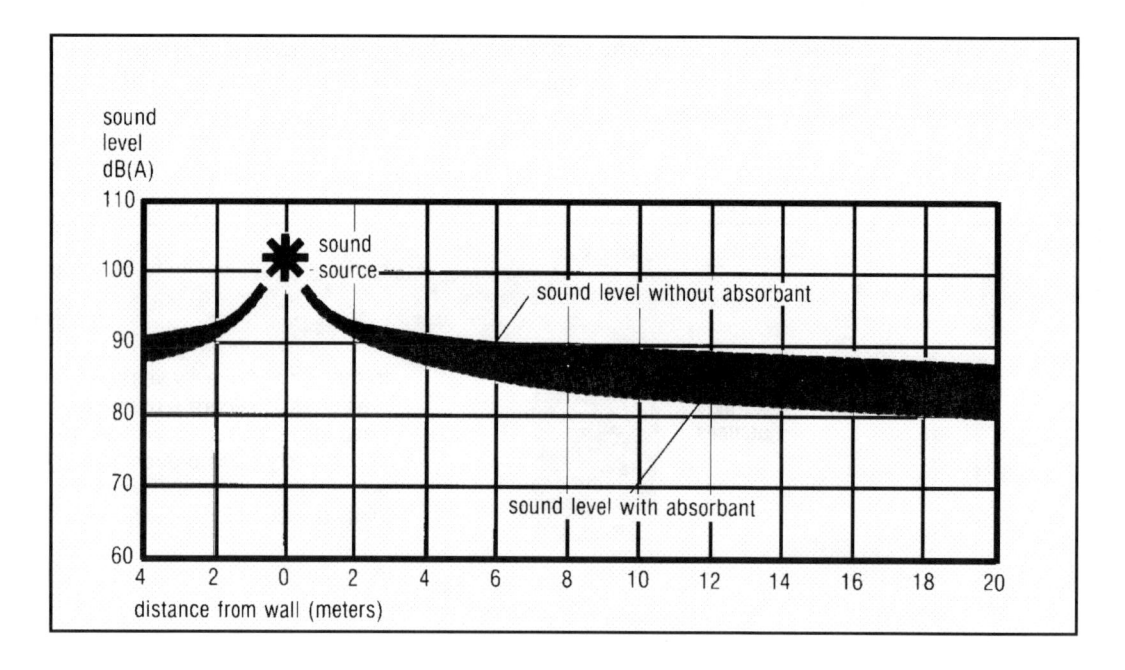

How sound levels vary at different distances from a sound surface before and after application of absorbent materials to the entire ceiling surface.

Sound Insulating Separate Rooms

With automation of machines and processes, remote control from a separate room may become desirable.

Some measures may include:

- Constructing the control rooms with materials having adequate TL
- Providing good sealing around doors and windows
- Providing openings for ventilation with passages for cables and piping equipped with good seals

The control room will need adequate ventilation and possibly air conditioning in hot working areas. Otherwise, there is risk that the doors will be opened for ventilation, which would drastically reduce the effectiveness of the room in reducing the noise level.

Example

Sound disturbances in an operating room or shop office may be caused by direct transmission (leakage through door openings, etc.) from the machine or by radiation from the common floor.

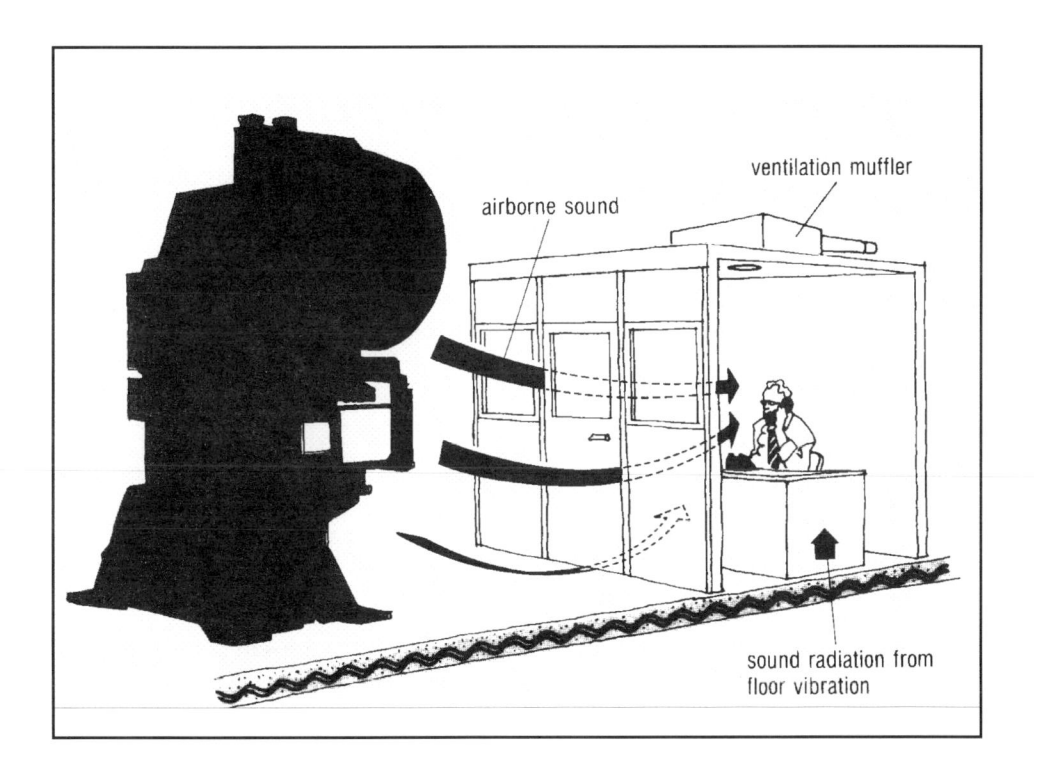

Maintenance

In some cases, a noise hazard will be created or made worse by a lack of maintenance. Parts may become loose, creating more noise because of improper operation or scraping against other parts. Grinding noises may also occur as the result of inadequate lubrication.

It is especially important to provide proper maintenance of noise control devices that are added or built into machinery. If a muffler becomes loose or worn out, for example, it should be fixed or replaced as soon as possible. Proactive maintenance programs where all machines are inspected on a scheduled basis is strongly recommended.

Planning for Noise Control

Noise control should be taken into account from the beginning of the planning process for a new building:

- The frame, floor, and machine bases should be selected so that all sources of distrubance can be provided withy effective vibration isolation.

Heavy equipment requires rigid, heavy bases. It is also possible to prevent the machine bases from making direct contact with the rest of the building frame.

- Whenever possible, select or specify machinery with low noise levels.

- Important noise sources may be surrounded with constructions which provide sound isolation. Special attention should be paid to portholes, observation windows, and other building parts which involve a risk of sound leakage.

- Noisy areas where the workers must spend time should be provided with ceiling coverings (and wall coverings as well in the case of very high ceilings) to absorb incoming sound.

The sound absorption ability varies greatly for various materials, so the materials must be selected in relation to the specific frequencies of the noise. Good sound absorption properties can often be combined with good thermal insulation.

- Office areas should be separated by means of a joint from building areas where vibration-producing equipment is installed.

- Wall and floor constructions as well as windows, doors, etc., should be constructed to provide the necessary sound TL.

- Fastening of noisy equipment to light separate structures should always be avoided because vibration isolation requires a rigid base to be effective.

- For office areas and storerooms where there are many functions in the same area, ceiling surfaces which provide good sound absorption and floors covered with soft textile material may be needed.

Example

Examples of noise control measures which can be carried out in an industrial building to avoid the spread of noise.

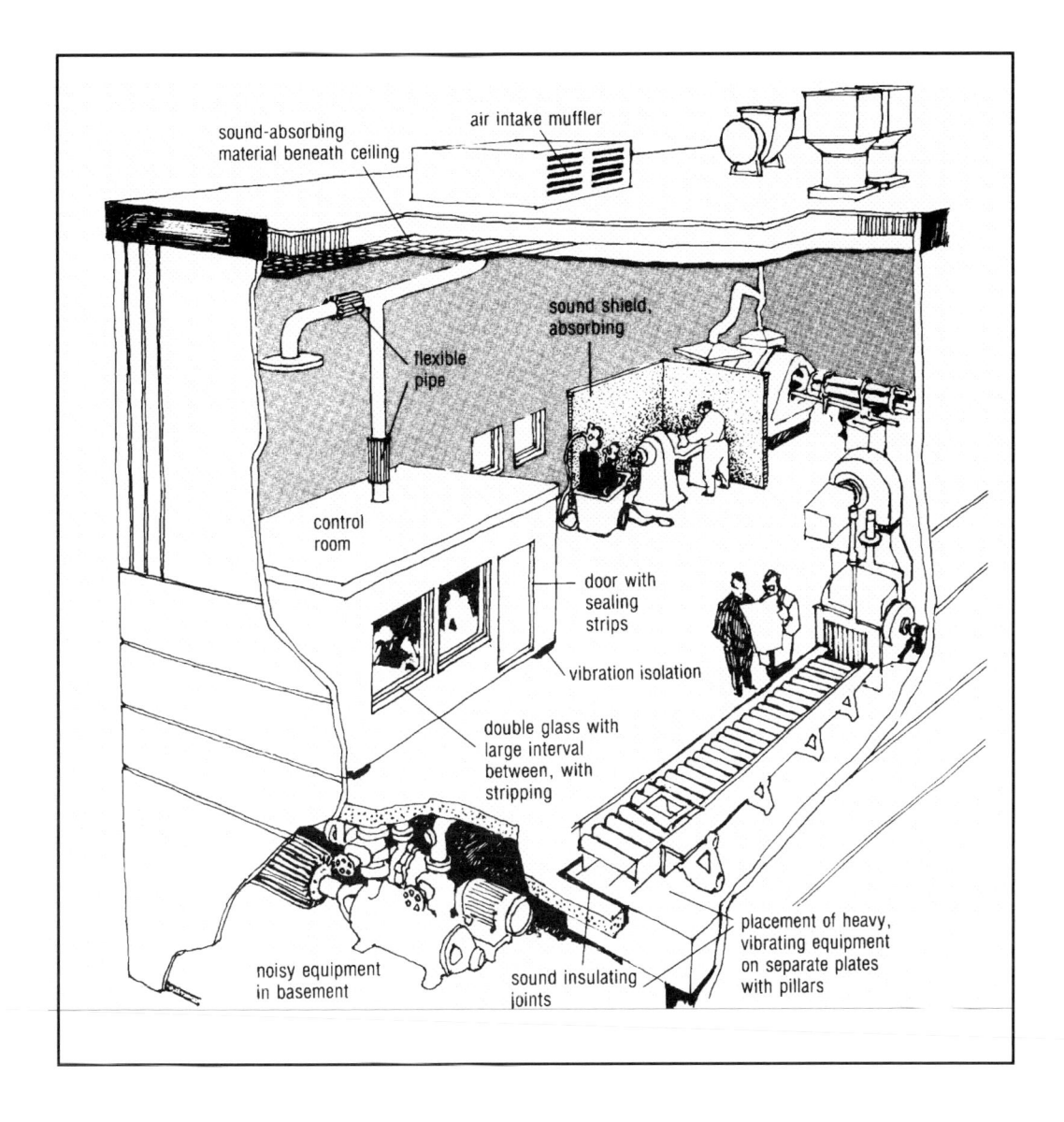

What is Required in the OSHA Standards?

Legal Limits on Noise

Under the Occupational Safety and Health Act, every employer is legally responsible for providing a workplace free of recognized hazards, such as excessive noise.

OSHA is generally the agency recognized with authority to enforce the noise control requirements specified in Title 29 of the Code of Federal Regulations, Part 1910 (General Industry), Subpart D (Occupational Health and Environmental Control), 1910.95 (Occupational Noise Exposure). However, approximately half of the states in the USA manage the OSHA functions as a state agency. Therefore, it is imperative that readers in state-plan states obtain and become familiar with their state-specific regulations. The rationale for this is that states must, at a minimum, have standards equivalent to the OSHA standards but also have the authority to implement more stringent requirements. Readers are also advised that this summary of OSHA requirements is neither a legal review nor a substitute for direct examination of the most current revision of 29 CFR 1910.95. For this reason, the reader should frequently consult the Federal Register (http://www.access.gpo.gov/nara/index.html) and the OSHA home page (http://www.osha.gov) on the internet.

The OSHA Occupational Noise standard requires that an employer whose employees are exposed to noise levels that equal or exceed an eight-hour time-weighted average (TWA) of 85 decibels on the "A" scale (slow response) or, equivalently, a daily dose of 50 percent, to administer a continuing, effective hearing conservation program. These levels are termed the Action Level.

Monitoring

An employer must ascertain if there are any areas or operations capable of exceeding the OSHA Action Level. Typically, the site health and safety professional utilizes a sound level meter and maps area noise levels. The sound level meter is calibrated before and after use and noise levels are documented on a site layout. This is often called a noise contour map.

A second approach to noise monitoring is the use of noise dosimeters. This approach is especially useful where circumstances such as high worker mobility, significant variations in sound levels, or a significant component of impulse noise make area monitoring generally inappropriate. Representative employees in areas that approach or exceed the 85 dBA Action Level are then equipped with calibrated dosimeters and full-shift noise levels are measured and documented. In this manner, the employer can clearly define areas and operations where the hearing conservation

What is Required in the OSHA Standards?

program needs to be implemented. The intent of the monitoring program is to identify employees for inclusion into the hearing conservation program. The program also assists in the identification of noise generating equipment. Such equipment then can be re-engineered to reduce noise levels. Proactive employers specify sound level limits for new equipment or specific engineering control measures to reduce area noise levels as described earlier in this guide.

The OSHA standard specifies limits on noise exposures in Table G-16 (Permissible Noise Exposures) of 29 CFR 1910.95 (*see:* p. 130). The following limits are specified:

Duration/day (hours)	Sound level (dBA, slow response)
8	90
6	92
4	95
3	97
2	100
1.5	102
1	105
0.5	110
0.25 or less	115

If noise levels cannot be reduced below those specified in the Permissible Noise Exposure standard, then the employer must select the proper types of hearing protectors for employees.

Noise monitoring is required to be repeated whenever a change in production, process, equipment, or controls increases noise exposure to the extent that additional employees may be exposed at or above the action level. The attenuation provided by hearing protectors being used by the employees may be rendered inadequate to meet the requirements listed in other parts of the regulation.

Employers are required to notify each employee who is exposed at or above an eight-hour time weighted average of 85 decibels of the results of the monitoring. Employees also have the right to observe noise measurements being conducted. Employees should be notified of the results and must be notified of those areas exceeding the action level.

Once a comprehensive noise measurement survey has been conducted, it need only be repeated when there has been change in production, equipment, or controls to the extent that additional employees may be exposed at or above the action level or the hearing protectors being used by employees are not adequate.

Audiometric Testing Program

For employees who work in areas that exceed the action level, annual audiometric evaluations must be performed. The Regulation specifies that the program shall be provided at no cost to employees and that specific criteria must be met in regards to who does testing, type of equipment, calibration procedures, and responsibility for the overall program. New employees must be tested within six months of first exposure at or above the action level.

Where industry utilizes mobile test van operations, the initial (baseline) audiogram can be obtained within one year of an employee's first exposure at or above the action level. Where baseline audiograms are obtained more than six months after the employee's first exposure at or above the action level, employees shall be required to wear hearing protectors for any period exceeding six months after first exposure until the baseline audiogram is obtained. Prior to testing a baseline evaluation, employees must be free from exposure to workplace noise at least 14 hours. The use of hearing protection can be utilized as part of the 14-hour period away from noise. In addition, employees should be notified of avoiding excessive noise exposures outside the workplace during the 14 hour quiet period.

Employees included in the audiometric monitoring must have their hearing checked annually and be compared to the baseline evaluation. If employees show a standard threshold shift as defined in 1910.95, or a professional reviewer such as an audiologist or other specialist has determined a need for further evaluation, then the employer shall refer the employee for that evaluation as appropriate. Referral can also be indicated where the employer suspects a medical pathology of the ear is caused or aggravated by the wearing of hearing protectors. The Regulation also indicates that the employee must be informed of test results and recommendations. The employer is also encouraged to review the use of hearing protection with employees, particularly if changes in hearing are determined to be a result of occupational noise exposures. Considerations and adjustments can also be made if an employee's hearing has improved.

Hearing Protectors

Employers must make hearing protectors available to all employees exposed to the action level of 85 dBA (TWA). Hearing protection is mandatory for any employee exposed to at least 90 dBA (TWA) or if hearing has shifted even if exposed to lesser levels of noise. It is the employer's responsibility to make sure that employees comply and to provide special training in the care and use of the protectors as well as provide proper sizing and fitting procedures for their employees. The protectors must attenuate the employee's exposure to at least 90 dBA over an eight-hour time-weighted average. If employees still exhibit changes in hearing as a result of occupational noise, then the protectors should attenuate employee exposure to levels of 85 dBA or below.

Training Program

The employer must provide a training program for all employees exposed to the action level. Emphasis should be placed on the intent of the Hearing Conservation Program as well as the effect of noise on hearing and how hearing loss can be prevented.

Further Information

More information about OSHA requirements can be obtained at the nearest federal or state OSHA office. They can provide you with copies of the current Regulation and will answer any specific questions regarding compliance with that Regulation. (See Appendices E and F for a listing of OSHA and state-plan offices.)

Consultation for Employers

In all states, free consultation services are available to help employers identify job safety and health hazards, such as excessive noise exposure, and to recommend solutions.

All information gathered by the consultant, including the employer's identity, is completely confidential and is not made available to OSHA enforcement personnel- with one very rare exception. If a consultant observes a *serious* violation of OSHA standards and the employer fails to correct it within the time period recommended by the consultant, OSHA or the responsible state agency will be notified.

Employers may on an anonymous basis contact the nearest OSHA office to find out how to take advantage of the free consultation services.

Loans for Small Businesses

The Small Business Administration (SBA), in cooperation with OSHA, provides loans to small businesses which are "likely to suffer substantial economic injury" in meeting OSHA safety and health standards.

Small business employers can obtain more information from the nearest SBA office or may write for a copy of *OSHA Handbook for Small Businesses* (OSHA 2209) from OSHA Office of Publications, P.O. Box 37535, Washington, D.C. 20201-7535, or via the Internet at: www.osha.gov.

Workers' Rights Under OSHA

With its small staff of inspectors, OSHA obviously cannot prevent every hazard in every workplace. It is essential that workers and employers cooperate on job safety and health programs.

In particular, OSHA encourages the formation of workplace health and safety committees. Health and safety committees can keep on top of job hazards such as noise, day in and day out. And if properly set up, they can make sure both employers and workers are involved in eliminating hazards. (For more information on health and safety committees and the workers' rights described below, contact the nearest OSHA office.)

Whenever possible, safety and health problems should be resolved at the workplace. If they can't be, here are some ways workers can use the laws:

- If your employer fails to correct a hazard causing an imminent danger, contact the nearest OSHA area office. If an OSHA inspector finds that an imminent danger exists, and your employer still fails to fix it, OSHA can seek an order from a federal court.

- If other types of hazards exist at your workplace, discuss them with other workers, your union representative, if any, and your employer. If you cannot get the hazard corrected, you can ask OSHA to make an inspection.

- OSHA will not tell your employer who requested an inspection if you ask that your name be kept confidential.

- During an OSHA inspection, a workers' representative may accompany the inspector, and must be paid by your employer.

- You (or your representative) may give OSHA information which could affect OSHA actions against your employer, such as proposed penalties and orders setting proposed dates for correcting hazards.

- If you disagree with the amount of time OSHA has given your employer to correct a hazard, you can ask for review by the Occupational Safety and Health Review Commission, an independent agency. If your employer asks the Commission to review OSHA actions, you can become a participant in the case by notifying the Commission. The address is Occupational Safety and Health Review Commission, 1120 20th Street, NW, 9th Floor, Washington, D.C. 20036-3457, or via the Internet at: www.OSHA.gov.

- You can obtain information about safety and health hazards, OSHA standards, and rights provided by OSHA!s law from your employer or from any OSHA office, listed in the back of this book or at www.OSHA.gov.

- Any employer with more than 10 employees must keep records of all work related injuries and illnesses, and you (or your representative) have the right to review those records.

- If there is a potential health hazard in Your workplace and you would like to know more about it, contact NIOSH-the National Institute for Occupational Safety and Health. NIOSH will give you information, and, if necessary, may conduct a survey of hazards in your workplace. If NIOSH has already studied hazards in your industry, it will give you the results. You can contact IOSH by telephone at: 1-800-35NIOSH (1-800-356-4674), or via the Internet at www.cdc.gov/NIOSH/homepage.html.

- Under section 18(b) of the Occupational Safety and Health Act of 1970, States are encouraged to develop and operate their own job safety and health plans. Currently there are 25 State plan States, of which 23 States administer plans covering both private and State and local government employment and two States cover only the public sector.

What is Required in the OSHA Standards?

OSHA closely monitors these State plans. States with approved programs must have standards that are identical to or at least as effective as the Federal standards, and must adopt new or reviewed Federal standards within six months of OSHA promulgation. Until such time as a State standard is promulgated, Federal OSHA will provide interim enforcement assistance, as appropriate, in these States. If you are not satisfied with any aspect of a State's performance, a Complaint About State Program Activities (CASPA) may be filed with Federal OSHA.

Workers' Right to Insist on Job Safety and Health

Section 11(c) of OSHA's law says an employer cannot punish or discriminate against an employee for job safety and health activities, such as:

- complaining to the employer, union, OSHA, or any other government agency about job safety and health hazards.

- filing safety or health grievances.

- participating in a workplace health and safety committee or union activities concerning job safety and health.

- participating in OSHA inspections, conferences, hearings, or other OSHA-related activities.

If you are exercising these or other safety and health rights, your employer cannot discriminate against you in any way, such as through firing, demotion, taking away seniority or benefits you've earned, transferring you to an undesirable job or shift, or threatening or harassing you.

If you believe you have been punished for using your safety and health rights, contact OSHA. You should do so *within 30 days* of the time you find out you've been discriminated against. If You want, your union representative can file your II(c) complaint for you.

If a state agency has been approved by OSHA to take over job safety and health enforcement, you may file your 11(c) complaint with either federal OSHA or the state agency.

OSHA Occupational Hearing Loss Recordkeeping

On July 1, 2002 OSHA published a final rule (*see:* Appendix C) on recordkeeping related to occupational hearing loss cases. The requirements for documenting hearing loss cases onto the OSHA Form 300 (Log of Work-Related Injuries and Illnesses) are explained in a question and answer format.

OSHA requires that employers with 11 or more employees record serious injuries and illnesses (generally those requiring treatment beyond first-aid) to maintain a log of work-related injuries and illnesses on the OSHA Form 300, with more detailed information on each case on an OSHA Form 301 (Injury and Illness Incident Report) or equivalent. The reader is referred to Title 29 Code of Federal Regulations, Part 1904 (Recording and Reporting Occupational Injuries and Illnesses) for details. These requirements as well as copies of the required OSHA recordkeeping forms can be accessed from the OSHA home page at http://www.osha.gov. Under the final rule OSHA published on recording hearing loss cases new criteria for recordability were detailed.

Under the new criteria, a hearing loss case becomes recordable if there is a shift in hearing in one or both ears of 25 decibels (dB) or more above audiometric zero, and there is a standard threshold shift (STS) of 10 dB in the same ear as the hearing loss. Age correction is permitted. (The reader is advised to review the latest revisions of 29 CFR 1904.10 on OSHA's website, www.osha.gov.) Any additional standard threshold shifts in the same ear would also be recordable. As in the prior standard, the employer may retest the employee within thirty days of the first hearing test showing the shift. The employer may also seek a medical opinion to determine if the hearing loss is work related.

How to Make Noise Measurements

Initial Screening Surveys

This type of survey (also called a "walk-through survey") may be quickly and inexpensively done with a basic sound level meter. Its purpose is to simply separate noise level exposures into three categories for each employee. Readings are taken at the employee's ear level:

High Noise Level

Where the noise level readings are almost continually over 85 dBA.

Low Noise Level

Where the noise level readings are almost continually below 85 dBA.

Borderline Noise Levels

Where noise levels fluctuate too much for an accurate level to be determined or if readings are found in a "borderline" area of approximately 82-88 dBA.

Note that while some sound level meters can only show the noise at that particular time, meters are available that give a time-weighted average.

Follow-Up Noise Surveys

Normally, a good portion of occupational noise exposures will clearly fall in either the "high" or "low" noise level categories, so that decisions concerning whether a Hearing Conservation Program is necessary can be readily made. Additional study must be made of "borderline" cases.

This can be done with a variety of equipment:

119

Noise Dosimeter

This instrument can be placed on an employee for an entire work shift, and then the "noise dose" can be calculated for that employee for that day. This instrument will provide fairly accurate measurements of variable noise levels and requires little attention other than at the beginning or end of the work shift. It usually provides the most accurate, long-term measurement of noise levels to which an employee is exposed, unless that employee is subjected to sharp, "impact" type noise. Always be sure to follow the manufacturer's instructions, set the dosimeter for an 80 dB cutoff (for an OSHA-type survey), and calibrate your equipment before and after each use.

Integrating Sound Level Meter

This is a unique noise measuring instrument that will provide an accurate measurement of variable, repetitive noise exposure in a short time. It is often used where employees use machinery on a noncontinuous basis on their job. For example, if you wanted to measure the noise exposure of a sewing machine operator, you would notice very quickly that the operator is exposed to two very different levels of noise the noise generated as the sewing machine is being run-"operating" noise, and the noise level as the employee is gathering material, trimming thread, arranging work, etc. This could be called "background" noise, because the noise exposure comes from other machines nearby rather than the employee's own machine.

Usually, this type of noise exposure is cyclic in nature—a period of time when the machine is operating and a period of time when it is not-and the cycles are usually repeated numerous times throughout the day.

Therefore, 10 or 20 minutes of a sample of noise exposure, if a number of repetitions are performed, may often be sufficient to accurately estimate the daily noise *if the job does not change.* Naturally, if the job does change, the sample time should be extended to as long a period as necessary to make certain that different noise levels and sources are included. Also, if the noise level is borderline, the sample time should be extended until an accurate determination of the daily noise dose can be made.

This type of instrument is also a popular one to use in "spotchecking" to see if noise levels recorded in previous studies are still present.

HOW TO SOLVE NOISE PROBLEMS

After you have determined that an employee is exposed to excessive noise and have involved that employee in your company's Hearing Conservation Program, attempts should be made to solve or eliminate the noise problem. Of course, if the noise is easily controllable or can be eliminated, this should be done immediately, but the protection of the hearing ability of the employee is of the greatest concern. So ear protection, training, and the other parts of a Hearing Conservation Program should be initiated immediately if engineering or noise control activity will involve more than an insignificant amount of time.

The noise solution/control process usually involves four steps:

1. Identification of the Noise Sources

When we speak of a noise source, we mean *exactly* where the noise is coming from. If it is a machine, your sound level meter or other instrument can be used to pinpoint the area of highest noise levels by moving the microphone in proximity to the machine itself. By this method, you can identify exactly whether it is machine operation noise, air ejection noise, air cooling noise, metal-to-metal contact, etc.

After you have isolated exactly what function on the machine appears to cause most of the noise, either shut that particular function off or remove that particular part, and, if possible, put the machine back in operation. This will tell you whether or not your solution of that one particular noise source will reduce the noise level sufficiently enough to enable it to be below allowable levels.

2. Identification of Central Noise Frequencies

In order to use the most cost-effective approach in controlling noise levels, it is often practical to obtain an Octave Band Frequency Analyzer to determine the exact frequencies in which most of the noise is being generated. In this way, you can compare this data with the technical data on various types of acoustical materials, damping materials, etc., to determine which one will be the most effective solution for your particular type of noise.

If you do not have access to an octave band frequency analyzer, and you do have a basic sound level meter with the A and C scales, you can at least determine whether there is a greater amount of high frequency noise than low frequency, or vice versa, at your particular noise source.

You would do this by holding the microphone of the instrument at the major source of noise on a piece of equipment you wish to control, and taking one set of readings on the A-scale. Then take another set of readings at the same location on the C-scale. If the C-scale readings are higher than the A-scale readings, the noise is generally better controlled with materials that absorb more low frequency noise than high frequency noise. The reverse would be true if the higher readings were on the A-scale or the readings were similar.

3. Installation of the Control Mechanism or Absorptive Material

The third step is the actual installation of the absorptive material or control device itself. There are two general schools of thought on this: first, use any material that you already have on hand. For example, if you are a carpet manufacturer, use carpet for your acoustical material. If you are a plywood manufacturer, use plywood. This may involve some trial and error, and you may end up having to use three or four times the acoustical material to accomplish the same noise reduction as you would with acoustical material that is especially manufactured for that purpose, but your cost will generally be significantly less.

The other approach would involve researching a number of catalogues and technical data to determine exactly what type of acoustical material best meets your needs as evidenced by your Octave Band noise analysis of your noise problem. The manufacturer's suggestion for installation and the ideas contained

What is Required in the OSHA Standards?

in this book will generally give satisfactory results if you have chosen your material properly.

4. *Recheck to Ensure Satisfactory Results*

A follow-up noise survey should be made to make sure that your acoustical treatment or control devices have satisfactorily reduced the amount of noise below the desired levels.

Appendix A

Sources of Information and Further Reading

Included with each reference is an Internet address which can be used by the reader who desires more in-depth information on noise and noise-control technologies.

E.H. Berger *et al.*, ed., *The Noise Manual,* 5th ed. Fairfax, Virginia: American Industrial Hygiene Association, 2000.

http://www.aiha.org

>This comprehensive reference presents information on noise control fundamentals, elements of a hearing conservation program, noise survey techniques, noise control engineering, hearing protection equipment, noise interference, and regulations.

Best's Safety & Security Directory. Millerton, NY: Grey House Publishing

http://www.greyhouse.com

>Published annually, this directory provides extensive information on vendors, OSHA Standards summaries, inspection checklists and training articles.

G.D. Clayton and F.E. Clayton, eds. *Patty's Industrial Hygiene and Toxicology,* 5th ed., vol 2, pt. III, *Physical Agents.* Hoboken, New Jersey: John Wiley & Sons, Inc., 2000.

http://www.wiley.com

>This volume presents information on the physics of sound, mechanism of hearing, acoustic trauma, effects of vibration, standards and exposure limits, hearing protection, noise control procedures and techniques, and community noise.

L.V. Cralley and L.J. Cralley, eds. *Industrial Hygiene Aspects of Plant Operations, vols. 1-3.* Caplan, New Jersey: Macmillan Publishing Co., 1982-1983.

Appendix A: Sources of Information and Further Reading

S.R. DiNardi. *The Occupational Environment—It's Evaluation and Control*. Fairfax, Virginia: American Industrial Hygiene Association.

http://www.aiha.org

This book presents information on the physics of sound, mechanisms of hearing, acoustic trauma, regulations, measurement of sound, noise control techniques, hearing conservation programs, and effects of vibration.

P.E. Hagan, J.F. Montgomery and J.T. O'Reilly, eds. *Accident Prevention Manual for Business and Industry—Administration & Programs,* 12th ed. Itasca, Illinois: National Safety Council, 2002.

P.E. Hagan, J.F. Montgomery and J.T. O'Reilly, eds. *Accident Prevention Manual for Business and Industry—Engineering & Technology,* 12th ed. Itasca, Illinois: National Safety Council, 2002.

http://www.nsc.org

The *Administration & Programs* volume presents information on how to design and implement safety programs in general. The *Engineering & Technology* volume presents information on noise control related to machine guards, metal casting, welding and cutting. It also includes a detailed overview of hearing protectors.

B.A. Plog, J. Niland, and P.J. Quinlan, eds. *Fundamentals of Industrial Hygiene,* 5th ed. Itasca, Illinois: National Safety Council, 2002.

http://www.nsc.org

This volume presents information on the physics of sound, acoustic trauma, mechanism of hearing, sound level measurement techniques, noise control techniques, hearing protection equipment, and an overview of regulations.

Appendix B

Government Agencies

Occupational Safety and Health Administration (OSHA)

http://www.osha.gov

The OSHA homepage on the Internet provides a bridge to an extensive library on noise and hearing conservation, including noise control techniques. The reader can use the Internet link provided above and insert the term "noise" into the search box on the OSHA homepage. Some of the reference links on the OSHA web site are:

- Work-Related Hearing Loss
- Health Hazard Evaluations: Noise and Hearing Loss, 1986-1997
- Occupational Noise - Induced Hearing Loss
- Sound Meter
- OSHA Technical Manual
- OSHA 3704, Hearing Conservation
- Hearing Conservation Program Checklist
- Preventing Occupational Hearing Loss - A Practical Guide
- The NIOSH Compendium of Hearing Protective Devices
- Industrial Noise Control Manual
- Best Practices in Hearing Loss Prevention
- OSHA Standards
- Criteria for a Recommended Standard: Occupational Noise Exposure   Revised Criteria
- Mine Safety and Health Administration - Health Standards for Occupational Noise Exposure
- Department of Defense Hearing Conservation Program (HCP) American National Standards Institute (ANSI)
- Health Hazard Evaluations (HHEs)

Appendix B: Government Agencies

> Noise Management in the Construction Industry: A Practical Approach
>
> Standard Interpretations and Compliance Letters

National Institutes for Occupational Safety and Health (NIOSH)

> http://www.cdc.gov/niosh/homepage.html
>
> The NIOSH homepage on the Internet also provides a bridge to an extensive library on noise and hearing conservation, including noise control techniques. The reader can use the Internet link provided above and insert the term "noise" into the search box on the NIOSH homepage. Some of the reference links on the NIOSH website are:
>
> Definition and Assessment of Engineering Noise Controls
>
> Noise and Hearing Loss Prevention
>
> Information and Resources
>
> Engineering Controls for Hearing Loss Prevention
>
> NIOSH Hearing Loss Publications
>
> Noise White Paper
>
> Standard Interpretations and Compliance Letters
>
> NIOSH - A Practical Guide to Preventing Hearing Loss
>
> Occupational Noise Exposure, Basis for the Exposure Standard
>
> Criteria for a Recommended Standard - Occupational Noise Exposure, 1998
>
> Investigation of Impulse Noise in Mining
>
> Carpenters' Noise Exposure,
>
> A Model Hearing Conservation Program for Coal Miners
>
> Best Practices in Hearing Loss Prevention
>
> Common Hearing Loss Prevention Terms
>
> Industrial Noise Control Manual
>
> Health Hazard Evaluations (HHES)
>
> NIOSH Health Care Worker Guidelines
>
> Investigation of Technology for Hearing Loss Prevention Control Technology Workshop (proceedings)
>
> NIOSH Compendium of Hearing Protective Devices Preventing Hearing Loss

Environmental Protection Agency (EPA)

http://www.epa.gov

The EPA homepage on the Internet provides an extensive listing of noise and noise control publications. The reader can use the Internet link above and type in the word "noise" in the search box on the EPA home page to access this extensive library. Some of the reference links are:

- Environmental Noise Control Act of 1972
- Noise Abatement and Control
- Noise Pollution
- Technology for Aircraft Noise Reduction
- Information on Levels of Environmental Noise Requisite to Protect
- Public Health and Welfare with an Adequate Margin of Safety
- Noise Source Abatement Technology and Cost Analysis Including Retrofitting

Mine Safety and Health Administration (MSHA)

ht@t :Hwww.msha.gov

The MSHA homepage on the Internet provides an extensive listing of noise and noise control 'ublications. The reader can use the Internet link above and type in the word "noise" to access this extensive library. Some of the reference links are:

- MSHA Health Standards for Occupational Noise Exposure Safety Standards for Underground Coal Mines
- Noise Regulations: Compliance Guide to MSHA's Occupational Noise Exposure Standard
- Noise Control Manual - Surface Mining
- Noise Control Manual - Preparation and Processing Plants

Appendix B: Government Agencies

Appendix C

CFR 1910.95—Occupational Noise Exposure

1910.95(a)

Protection against the effects of noise exposure shall be provided when the sound levels exceed those shown in Table G-16 (p. 130) when measured on the A scale of a standard sound level meter at slow response. When noise levels are determined by octave band analysis, the equivalent A-weighted sound level may be determined as follows:

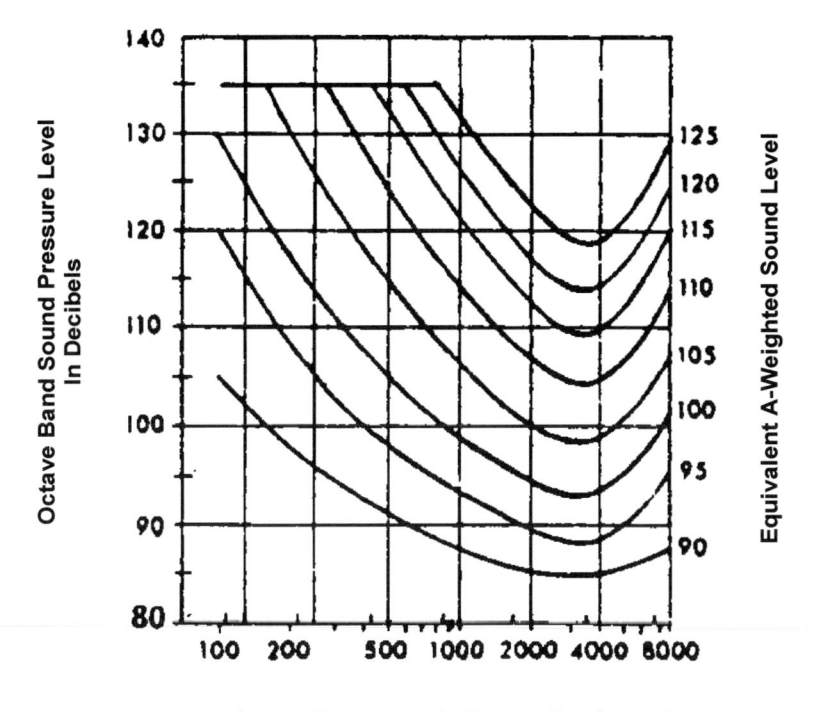

Figure G-9. Equivalent A-Weighted Sound Level

Appendix C: Occupational Noise Exposure

Equivalent sound level contours. Octave band sound pressure levels may be converted to the equivalent A-weighted sound level by plotting them on this graph and noting the A-weighted sound level corresponding to the point of highest penetration into the sound level contours. This equivalent A-weighted sound level, which may differ from the actual A-weighted sound level of the noise, is used to determine exposure limits from Table G-16.

Table G-16—Permissible Noise Exposures [1]

Duration per day hours	Sound level dBA slow response
8	90
6	92
4	95
3	97
2	100
1-1/2	102
1	105
1/2	110
1/4 or less	115

[1]When the daily noise exposure is composed of two or more periods of noise exposure of different levels, their combined effect should be considered, rather than the individual effect of each. If the sum of the following fractions: $C(1)/T(1) + C(2)/T(2)C(n)/T(n)$ exceeds unity, then, the mixed exposure should be considered to exceed the limit value. $C(n)$ indicates the total time of exposure at a specified noise level, and $T(n)$ indicates the total time of exposure permitted at that level. Exposure to impulsive or impact noise should not exceed 140 dB peak sound pressure level.

..1910.95(c)(1)

1910.95(b)

1910.95(b)(1)
When employees are subjected to sound exceeding those listed in Table G-16, feasible administrative or engineering controls shall be utilized. If such controls fail to reduce sound levels within the levels of Table G-16, personal protective equipment shall be provided and used to reduce sound levels within the levels of the table.

1910.95(b)(2)
If the variations in noise level involve maxima at intervals of 1 second or less, it is to be considered continuous.

1910.95(b)

1910.95(b)(1)

When employees are subjected to sound exceeding those listed in Table G-16, feasible administrative or engineering controls shall be utilized. If such controls fail to reduce sound levels within the levels of Table G-16, personal protective equipment shall be provided and used to reduce sound levels within the levels of the table.

1910.95(b)(2)

If the variations in noise level involve maxima at intervals of 1 second or less, it is to be considered continuous.

1910.95(c)

1910.95(c)

"Hearing conservation program."

1910.95(c)(1)

The employer shall administer a continuing, effective hearing conservation program, as described in paragraphs (c) through (o) of this section, whenever employee noise exposures equal or exceed an 8-hour time-weighted average sound level (TWA) of 85 decibels measured on the A scale (slow response) or, equivalently, a dose of fifty percent. For purposes of the hearing conservation program, employee noise exposures shall be computed in accordance with appendix A and Table G-16A [*see:* pgs. 143-44], and without regard to any attenuation provided by the use of personal protective equipment.

1910.95(c)(2)

For purposes of paragraphs (c) through (n) of this section, an 8-hour time-weighted average of 85 decibels or a dose of fifty percent shall also be referred to as the action level.

1910.95(d)

"Monitoring."

1910.95(d)(1)

When information indicates that any employee's exposure may equal or exceed an 8-hour time-weighted average of 85 decibels, the employer shall develop and implement a monitoring program.

1910.95(d)(1)(i)

The sampling strategy shall be designed to identify employees for inclusion in the hearing conservation program and to enable the proper selection of hearing

Appendix C: Occupational Noise Exposure

protectors.

1910.95(d)(1)(ii)

Where circumstances such as high worker mobility, significant variations in sound level, or a significant component of impulse noise make area monitoring generally inappropriate, the employer shall use representative personal sampling to comply with the monitoring requirements of this paragraph unless the employer can show that area sampling produces equivalent results.

1910.95(d)(2)

1910.95(d)(2)

1910.95(d)(2)(i)

All continuous, intermittent and impulsive sound levels from 80 decibels to 130 decibels shall be integrated into the noise measurements.

1910.95(d)(2)(ii)

Instruments used to measure employee noise exposure shall be calibrated to ensure measurement accuracy.

1910.95(d)(3)

Monitoring shall be repeated whenever a change in production, process, equipment or controls increases noise exposures to the extent that:

1910.95(d)(3)(i)

Additional employees may be exposed at or above the action level; or 1910.95(d)(3)(ii) The attenuation provided by hearing protectors being used by employees may be rendered inadequate to meet the requirements of paragraph (j) of this section.

1910.95(e)

"Employee notification." The employer shall notify each employee exposed at or above an 8-hour time-weighted average of 85 decibels of the results of the monitoring.

1910.95(f)

"Observation of monitoring." The employer shall provide affected employees or their representatives with an opportunity to observe any noise measurements conducted pursuant to this section.

..1910.95(g)

1910.95(g)

"Audiometric testing program."

1910.95(g)(1)

The employer shall establish and maintain an audiometric testing program as provided in this paragraph by making audiometric testing available to all employees whose exposures equal or exceed an 8-hour time-weighted average of 85 decibels.

1910.95(g)(2)

The program shall be provided at no cost to employees.

1910.95(g)(3)

Audiometric tests shall be performed by a licensed or certified audiologist, otolaryngology's, or other physician, or by a technician who is certified by the Council of Accreditation in Occupational Hearing Conservation, or who has satisfactorily demonstrated competence in administering audiometric examinations, obtaining valid audiograms, and properly using, maintaining and checking calibration and proper functioning of the audiometers being used. A technician who operates microprocessor audiometers does not need to be certified. A technician who performs audiometric tests must be responsible to an audiologist, otolaryngologist or physician.

1910.95(g)(4)

All audiograms obtained pursuant to this section shall meet the requirements of Appendix C: "Audiometric Measuring Instruments."

1910.95(g)(5)

"Baseline audiogram."

1910.95(g)(5)(i)

Within 6 months of an employee's first exposure at or above the action level, the employer shall establish a valid baseline audiogram against which subsequent audiograms can be compared.

..1910.95(g)(5)(ii)

1910.95(g)(5)(ii)

"Mobile test van exception." Where mobile test vans are used to meet the audiometric testing obligation, the employer shall obtain a valid baseline audiogram within 1 year of an employee's first exposure at or above the action level. Where baseline audiograms are obtained more than 6 months after the employee's first exposure at or above the action level, employees shall wear hearing protectors for any period exceeding six months after first exposure until the baseline audiogram is obtained.

1910.95(g)(5)(iii)

Testing to establish a baseline audiogram shall be preceded by at least 14 hours without exposure to workplace noise. Hearing protectors may be used as a substitute for the requirement that baseline audiograms be preceded by 14 hours without exposure to workplace noise.

Appendix C: Occupational Noise Exposure

1910.95(g)(5)(iv)

The employer shall notify employees of the need to avoid high levels of non-occupational noise exposure during the 14-hour period immediately preceding the audiometric examination.

1910.95(g)(6)

"Annual audiogram." At least annually after obtaining the baseline audiogram, the employer shall obtain a new audiogram for each employee exposed at or above an 8-hour time-weighted average of 85 decibels.

1910.95(g)(7)

"Evaluation of audiogram."

1910.95(g)(7)(i)

Each employee's annual audiogram shall be compared to that employee's baseline audiogram to determine if the audiogram is valid and if a standard threshold shift as defined in paragraph (g)(10) of this section has occurred. This comparison may be done by a technician.

..1910.95(g)(7)(ii)

1910.95(g)(7)(ii)

If the annual audiogram shows that an employee has suffered a standard threshold shift, the employer may obtain a retest within 30 days and consider the results of the retest as the annual audiogram.

1910.95(g)(7)(iii)

The audiologist, otolaryngologist, or physician shall review problem audiograms and shall determine whether there is a need for further evaluation. The employer shall provide to the person performing this evaluation the following information:

1910.95(g)(7)(iii)(A)

A copy of the requirements for hearing conservation as set forth in paragraphs (c) through (n) of this section;

1910.95(g)(7)(iii)(B)

The baseline audiogram and most recent audiogram of the employee to be evaluated;

1910.95(g)(7)(iii)(C)

Measurements of background sound pressure levels in the audiometric test room as required in Appendix D: Audiometric Test Rooms.

1910.95(g)(7)(iii)(D)

Records of audiometer calibrations required by paragraph (h)(5) of this section.

..1910.95(g)(8)

1910.95(g)(8)

"Follow-up procedures."

1910.95(g)(8)(i)

If a comparison of the annual audiogram to the baseline audiogram indicates a standard threshold shift as defined in paragraph (g)(10) of this section has occurred, the employee shall be informed of this fact in writing, within 21 days of the determination.

1910.95(g)(8)(ii)

Unless a physician determines that the standard threshold shift is not work related or aggravated by occupational noise exposure, the employer shall ensure that the following steps are taken when a standard threshold shift occurs:

1910.95(g)(8)(ii)(A)

Employees not using hearing protectors shall be fitted with hearing protectors, trained in their use and care, and required to use them.

1910.95(g)(8)(ii)(B)

Employees already using hearing protectors shall be refitted and retrained in the use of hearing protectors and provided with hearing protectors offering greater attenuation if necessary.

1910.95(g)(8)(ii)(C)

The employee shall be referred for a clinical audiological evaluation or an otological examination, as appropriate, if additional testing is necessary or if the employer suspects that a medical pathology of the ear is caused or aggravated by the wearing of hearing protectors.

1910.95(g)(8)(ii)(D)

The employee is informed of the need for an otological examination if a medical pathology of the ear that is unrelated to the use of hearing protectors is suspected.

..1910.95(g)(8)(iii)

1910.95(g)(8)(iii)

If subsequent audiometric testing of an employee whose exposure to noise is less than an 8-hour TWA of 90 decibels indicates that a standard threshold shift is not persistent, the employer:

1910.95(g)(8)(iii)(A)

Shall inform the employee of the new audiometric interpretation; and

Appendix C: Occupational Noise Exposure

1910.95(g)(8)(iii)(B)

May discontinue the required use of hearing protectors for that employee.

1910.95(g)(9)

"Revised baseline." An annual audiogram may be substituted for the baseline audiogram when, in the judgment of the audiologist, otolaryngologist or physician who is evaluating the audiogram:1910.95(g)(9)(i)

The standard threshold shift revealed by the audiogram is persistent; or

1910.95(g)(9)(ii).

The hearing threshold shown in the annual audiogram indicates significant improvement over the baseline audiogram.

1910.95(g)(10)

"Standard threshold shift."

1910.95(g)(10)(i)

As used in this section, a standard threshold shift is a change in hearing threshold relative to the baseline audiogram of an average of 10 dB or more at 2000, 3000, and 4000 Hz in either ear.

..1910.95(g)(10)(ii)

1910.95(g)(10)(ii)

In determining whether a standard threshold shift has occurred, allowance may be made for the contribution of aging (presbycusis) to the change in hearing level by correcting the annual audiogram according to the procedure described in Appendix F: "Calculation and Application of Age Correction to Audiograms."

1910.95(h)

"Audiometric test requirements."

1910.95(h)(1)

Audiometric tests shall be pure tone, air conduction, hearing threshold examinations, with test frequencies including as a minimum 500, 1000, 2000, 3000, 4000, and 6000 Hz. Tests at each frequency shall be taken separately for each ear.

1910.95(h)(2)

Audiometric tests shall be conducted with audiometers (including microprocessor audiometers) that meet the specifications of, and are maintained and used in accordance with, American National Standard Specification for Audiometers, S3.6-1969, which is incorporated by reference as specified in Sec. 1910.6.

1910.95(h)(3)

Pulsed-tone and self-recording audiometers, if used, shall meet the requirements specified in Appendix C: "Audiometric Measuring Instruments."

1910.95(h)(4)

Audiometric examinations shall be administered in a room meeting the requirements listed in Appendix D: "Audiometric Test Rooms."

..1910.95(h)(5)

1910.95(h)(5)

"Audiometer calibration."

1910.95(h)(5)(i)

The functional operation of the audiometer shall be checked before each day's use by testing a person with known, stable hearing thresholds, and by listening to the audiometer's output to make sure that the output is free from distorted or unwanted sounds. Deviations of 10 decibels or greater require an acoustic calibration.

1910.95(h)(5)(ii)

Audiometer calibration shall be checked acoustically at least annually in accordance with Appendix E: "Acoustic Calibration of Audiometers." Test frequencies below 500 Hz and above 6000 Hz may be omitted from this check. Deviations of 15 decibels or greater require an exhaustive calibration.

1910.95(h)(5)(iii)

An exhaustive calibration shall be performed at least every two years in accordance with sections 4.1.2; 4.1.3.; 4.1.4.3; 4.2; 4.4.1; 4.4.2; 4.4.3; and 4.5 of the American National Standard Specification for Audiometers, S3.6-1969. Test frequencies below 500 Hz and above 6000 Hz may be omitted from this calibration.

1910.95(i)

"Hearing protectors."

1910.95(i)(1)

Employers shall make hearing protectors available to all employees exposed to an 8-hour time-weighted average of 85 decibels or greater at no cost to the employees. Hearing protectors shall be replaced as necessary.

1910.95(i)(2)

Employers shall ensure that hearing protectors are worn:

1910.95(i)(2)(i)

By an employee who is required by paragraph (b)(1) of this section to wear personal protective equipment; and

Appendix C: Occupational Noise Exposure

..1910.95(i)(2)(ii)

1910.95(i)(2)(ii)

By any employee who is exposed to an 8-hour time-weighted average of 85 decibels or greater, and who:

1910.95(i)(2)(ii)(A)

Has not yet had a baseline audiogram established pursuant to paragraph (g)(5)(ii); or

1910.95(i)(2)(ii)(B)

Has experienced a standard threshold shift.

1910.95(i)(3)

Employees shall be given the opportunity to select their hearing protectors from a variety of suitable hearing protectors provided by the employer.

1910.95(i)(4)

The employer shall provide training in the use and care of all hearing protectors provided to employees.

1910.95(i)(5)

The employer shall ensure proper initial fitting and supervise the correct use of all hearing protectors.

1910.95(j)

"Hearing protector attenuation."

1910.95(j)(1)

The employer shall evaluate hearing protector attenuation for the specific noise environments in which the protector will be used. The employer shall use one of the evaluation methods described in Appendix B: "Methods for Estimating the Adequacy of Hearing Protection Attenuation."

..1910.95(j)(2)

1910.95(j)(2)

Hearing protectors must attenuate employee exposure at least to an 8-hour time-weighted average of 90 decibels as required by paragraph (b) of this section.

1910.95(j)(3)

For employees who have experienced a standard threshold shift, hearing protectors must attenuate employee exposure to an 8-hour time-weighted average of 85 decibels or below.

1910.95(j)(4)

The adequacy of hearing protector attenuation shall be re-evaluated whenever employee noise exposures increase to the extent that the hearing protectors provided may no longer provide adequate attenuation. The employer shall provide more effective hearing protectors where necessary.

1910.95(k)

"Training program."

1910.95(k)(1)

The employer shall institute a training program for all employees who are exposed to noise at or above an 8-hour time-weighted average of 85 decibels, and shall ensure employee participation in such program.

1910.95(k)(2)

The training program shall be repeated annually for each employee included in the hearing conservation program. Information provided in the training program shall be updated to be consistent with changes in protective equipment and work processes.

1910.95(k)(3)

The employer shall ensure that each employee is informed of the following:

..*1910.95(k)(3)(i)*

1910.95(k)(3)(i)

The effects of noise on hearing;

1910.95(k)(3)(ii)

The purpose of hearing protectors, the advantages, disadvantages, and attenuation of various types, and instructions on selection, fitting, use, and care; and

1910.95(k)(3)(iii)

The purpose of audiometric testing, and an explanation of the test procedures.

1910.95(l)

"Access to information and training materials."

1910.95(l)(1)

The employer shall make available to affected employees or their representatives copies of this standard and shall also post a copy in the workplace.

1910.95(l)(2)

The employer shall provide to affected employees any informational materials pertaining to the standard that are supplied to the employer by the Assistant Secretary.

Appendix C: Occupational Noise Exposure

1910.95(l)(3)

The employer shall provide, upon request, all materials related to the employer's training and education program pertaining to this standard to the Assistant Secretary and the Director.

..1910.95(m)

1910.95(m)

"Recordkeeping"

1910.95(m)(1)

"Exposure measurements." The employer shall maintain an accurate record of all employee exposure measurements required by paragraph (d) of this section.

1910.95(m)(2)

"Audiometric tests."

1910.95(m)(2)(i)

The employer shall retain all employee audiometric test records obtained pursuant to paragraph (g) of this section:

1910.95(m)(2)(ii)

This record shall include:

1910.95(m)(2)(ii)(A)

Name and job classification of the employee;

1910.95(m)(2)(ii)(B)

Date of the audiogram;

1910.95(m)(2)(ii)(C)

The examiner's name;

1910.95(m)(2)(ii)(D)

Date of the last acoustic or exhaustive calibration of the audiometer; and

1910.95(m)(2)(ii)(E)

Employee's most recent noise exposure assessment.

1910.95(m)(2)(ii)(F)

The employer shall maintain accurate records of the measurements of the background sound pressure levels in audiometric test rooms.

1910.95(m)(3)Record retention. The employer shall retain records required in this paragraph (m) for at least the following periods.

..1910.95(m)(3)(i)

1910.95(m)(3)(i)

Noise exposure measurement records shall be retained for two years.

1910.95(m)(3)(ii)

Audiometric test records shall be retained for the duration of the affected employee's employment.

1910.95(m)(4)

"Access to records." All records required by this section shall be provided upon request to employees, former employees, representatives designated by the individual employee, and the Assistant Secretary. The provisions of 29 CFR 1910.20 (a)-(e) and (g)-

1910.95(m)(4)(i)

apply to access to records under this section.

1910.95(m)(5)

"Transfer of records." If the employer ceases to do business, the employer shall transfer to the successor employer all records required to be maintained by this section, and the successor employer shall retain them for the remainder of the period prescribed in paragraph (m)(3) of this section.

1910.95(n)

"Appendices."

1910.95(n)(1)

Appendices A, B, C, D, and E to this section are incorporated as part of this section and the contents of these appendices are mandatory.

..1910.95(n)(2)

1910.95(n)(2)

Appendices F and G to this section are informational and are not intended to create any additional obligations not otherwise imposed or to detract from any existing obligations.

1910.95(o)

"Exemptions." Paragraphs (c) through (n) of this section shall not apply to employers engaged in oil and gas well drilling and servicing operations.

Appendix C: Occupational Noise Exposure

1910.95(p)

"Startup date." Baseline audiograms required by paragraph (g) of this section shall be completed by March 1, 1984.

[39 FR 23502, June 27, 1974, as amended at 46 FR 4161, Jan. 16, 1981; 46 FR 62845, Dec. 29, 1981; 48 FR 9776, Mar. 8, 1983; 48 FR 29687, June 28, 1983; 54 FR 24333, June 7, 1989; 61 FR 5507, Feb. 13, 1996; 61 FR 9227, March 7, 1996]

1910.95 Appendix A

Noise Exposure Computation

This Appendix is Mandatory

I. Computation of Employee Noise Exposure

(1) Noise dose is computed using Table G-16A as follows:

(i) When the sound level, L, is constant over the entire work shift, the noise dose, D, in percent, is given by: D=100 C/T where C is the total length of the work day, in hours, and T is the reference duration corresponding to the measured sound level, L, as given in Table G-16A or by the formula shown as a footnote to that table.

(ii) When the workshift noise exposure is composed of two or more periods of noise at different levels, the total noise dose over the work day is given by:

$$D = 100 \ (C(1)/T(1) + C(2)/T(2) + ... + C(n)/T(n)),$$

where C(n) indicates the total time of exposure at a specific noise level, and T(n) indicates the reference duration for that level as given by Table G-16A.

(2) The eight-hour time-weighted average sound level (TWA), in decibels, may be computed from the dose, in percent, by means of the formula:

$$TWA = 16.61 \log(10) \ (D/100) + 90.$$

For an eight-hour workshift with the noise level constant over the entire shift, the TWA is equal to the measured sound level.

(3) A table relating dose and TWA is given in Section II.

Table G-16A.

A-weighted sound level, L (decibel)	Reference duration, T (hour)
80	32
81	27.9
82	24.3
83	21.1
84	18.4
85	16
86	13.9
87	12.1
88	10.6
89	9.2
90	8
91	.0
92	6.1
93	5.3
94	4.6
95	4
96	3.5
97	3.0
98	2.6
99	2.3
100	2
101	1.7
102	1.5
103	1.3
104	1.1
105	1
106	0.87
107	0.76
108	0.66
109	0.57
110	0.5
111	0.44

(Continued on page 144)

Appendix C: Occupational Noise Exposure

(Continued from page 143)

Table G-16A—Continued

A-weighted sound level, L (decibel)	Reference duration, T (hour)
112	0.38
113	0.33
114	0.29
115	0.25
116	0.22
117	0.19
118	0.16
119	0.14
120	0.125
121	0.11
122	0.095
123	0.082
124	0.072
125	0.063
126	0.054
127	0.047
128	0.041
129	0.036
130	0.031

In the above table the reference duration, T, is computed by

$$T = \frac{8}{2^{(L-90)/5}}$$

where L is the measured A-weighted sound level.

II. Conversion Between "Dose" and "8-Hour Time-Weighted Average"

Sound Level

Compliance with paragraphs (c)-(r) of this regulation is determined by the amount of exposure to noise in the workplace. The amount of such exposure is usually measured with an audiodosimeter which gives a readout in terms of "dose." In order to better understand the requirements of the amendment, dosimeter readings can be converted to an "8-hour time-weighted average sound level." (TWA).In order to convert the reading of a dosimeter into TWA, see Table A-1, below. This table applies to dosimeters that are set by the manufacturer to calculate dose or percent exposure according to the relationships in Table G-16A. So, for example, a dose of 91 percent over an eight hour day results in a TWA of 89.3 dB, and, a dose of 50 percent corresponds to a TWA of 85 dB.

If the dose as read on the dosimeter is less than or greater than the values found in Table A-1, the TWA may be calculated by using the formula: TWA = 16.61 log(10) (D/100) + 90 where TWA=8-hour time-weighted average sound level and D = accumulated dose in percent exposure.

Table A-1. Conversion from "Percent Noise Exposure" or "Dose" to "8-Hour Time-Weighted Average Sound Level" (TWA)

Dose or percent noise exposure	TWA
10	73.4
15	76.3
20	78.4
25	80.0
30	1.3
35	82.4
40	83.4
45	84.2
50	85.0
55	85.7
60	86.3
65	86.9
70	87.4
75	87.9
80	88.4
81	88.5

(Continued on page 146)

Appendix C: Occupational Noise Exposure

(Continued from page 145)

Table A-1—Continued

<u>Dose or percent noise exposure</u>   <u>TWA</u>

Dose	TWA
82	88.6
83	88.7
84	88.7
85	88.8
86	88.9
87	89.0
88	89.1
89	89.2
90	89.2
91	89.3
92	89.4
93	89.5
94	89.6
95	89.6
96	89.7
97	89.8
98	89.9
99	89.9
100	90.0
101	90.1
102	90.1
103	90.2
104	90.3
105	90.4
106	90.4
107	90.5
108	90.6
109	0.6
110	90.7
111	90.8
112	90.8
113	90.9
114	90.9
115	91.1
116	91.1
117	91.1
118	1.2
119	91.3
120	91.3
125	91.6
130	91.9

(Continued on page 147)

(Continued from page 146)

Table A-1—Continued

Dose or percent noise exposure	TWA
135	92.2
140	92.4
145	92.7
150	92.9
155	93.2
160	93.4
165	93.6
170	93.8
175	94.0
180	94.2
185	94.4
190	94.6
195	94.8
200	95.0
210	95.4
220	95.7
230	96.0
240	96.3
250	96.6
260	96.9
270	97.2
280	97.4
290	97.7
300	97.9
310	98.2
320	98.4
330	98.6
340	98.8
350	99.0
360	99.2
370	99.4
380	99.6
390	99.8
400	100.0
410	100.2
420	100.4
430	100.5
440	100.7
450	100.8
460	101.0
470	101.2

(Continued on page 148)

Appendix C: Occupational Noise Exposure

(Continued from page 147)

Table A-1—Continued

Dose or percent noise exposure	TWA
480	101.3
490	101.5
500	101.6
510	101.8
520	101.9
530	102.0
540	102.2
550	102.3
560	102.4
570	102.6
580	102.7
590	102.8
600	102.9
610	103.0
620	103.2
630	103.3
640	103.4
650	103.5
660	103.6
670	103.7
680	103.8
690	103.9
700	104.0
710	104.1
720	104.2
730	104.3
740	104.4
750	104.5
760	104.6
770	104.7
780	104.8
790	104.9
800	105.0
810	105.1
820	105.2
830	105.3
840	105.4
850	105.4
860	105.5
870	105.6

(Continued on page 149)

(Continued from page 148)

Table A-1—Continued

Dose or percent noise exposure	TWA
880	105.7
890	105.8
900	105.8
910	105.9
920	106.0
930	106.1
940	106.2
950	106.2
960	106.3
970	106.4
980	106.5
990	106.5
999	106.6

1910.95 Appendix B—Methods for Estimating the Adequacy of Hearing Protector Attenuation

1910.95 App B U.S.
Department of Labor
Occupational Safety & Health Administration

Part: Number:	1910
Part Title:	Occupational Safety and Health Standards
Subpart:	G
Subpart Title:	Occupational Health and Environment Control
Standard Number:1	910.95 App B
Title:	Methods for estimating the adequacy of hearing protector attenuation

This Appendix is Mandatory

For employees who have experienced a significant threshold shift, hearing protector attenuation must be sufficient to reduce employee exposure to a TWA of 85 dB. Employers must select one of the following methods by which to estimate the adequacy of hearing protector attenuation.

The most convenient method is the Noise Reduction Rating (NRR) developed by the Environmental Protection Agency (EPA). According to EPA regulation, the NRR

Appendix C: Occupational Noise Exposure

must be shown on the hearing protector package. The NRR is then related to an individual worker's noise environment in order to assess the adequacy of the attenuation of a given hearing protector. This appendix describes four methods of using the NRR to determine whether a particular hearing protector provides adequate protection within a given exposure environment. Selection among the four procedures is dependent upon the employer's noise measuring instruments.

Instead of using the NRR, employers may evaluate the adequacy of hearing protector attenuation by using one of the three methods developed by the National Institute for Occupational Safety and Health (NIOSH), which are described in the "List of Personal Hearing Protectors and Attenuation Data," HEW Publication No. 76-120, 1975, pages 21-37. These methods are known as NIOSH methods No. 1, No. 2 and No. 3. The NRR described below is a simplification of NIOSH method No. 2. The most complex method is NIOSH method No. 1, which is probably the most accurate method since it uses the largest amount of spectral information from the individual employee's noise environment. As in the case of the NRR method described below, if one of the NIOSH methods is used, the selected method must be applied to an individual's noise environment to assess the adequacy of the attenuation. Employers should be careful to take a sufficient number of measurements in order to achieve a representative sample for each time segment.

NOTE: The employer must remember that calculated attenuation values reflect realistic values only to the extent that the protectors are properly fitted and worn. When using the NRR to assess hearing protector adequacy, one of the following methods must be used:

(i) When using a dosimeter that is capable of C-weighted measurements:

(A) Obtain the employee's C-weighted dose for the entire workshift, and convert to TWA (see appendix A, II).

(B) Subtract the NRR from the C-weighted TWA to obtain the estimated A-weighted TWA under the ear protector.

(ii) When using a dosimeter that is not capable of C-weighted measurements, the following method may be used:

(A) Convert the A-weighted dose to TWA (see appendix A).

(B) (B) Subtract 7 dB from the NRR.

(C) Subtract the remainder from the A-weighted TWA to obtain the estimated A-weighted TWA under the ear protector.

(iii) When using a sound level meter set to the A-weighting network:

(A) Obtain the employee's A-weighted TWA.

(B) Subtract 7 dB from the NRR, and subtract the remainder from the A-weighted TWA to obtain the estimated A-weighted TWA under the ear protector.

(iv) When using a sound level meter set on the C-weighting network:

(A) Obtain a representative sample of the C-weighted sound levels in the employee's environment.

(B) Subtract the NRR from the C-weighted average sound level to obtain the estimated A-weighted TWA under the ear protector.

(v) When using area monitoring procedures and a sound level meter set to the A-weighing network.

(A) Obtain a representative sound level for the area in question.

(B) Subtract 7 dB from the NRR and subtract the remainder from the A-weighted sound level for that area.

(vi) When using area monitoring procedures and a sound level meter set to the C-weighting network:

(A) Obtain a representative sound level for the area in question.

(B) Subtract the NRR from the C-weighted sound level for that area.

1910.95 Appendix C—Audiometric Measuring Instruments

Part Number:	1910
Part Title:	Occupational Safety and Health Standards
Subpart:	G
Subpart Title:	Occupational Health and Environment Control
Standard Number:	1910.95 App C
Title:	Audiometric measuring instruments

This Appendix is Mandatory

1. In the event that pulsed-tone audiometers are used, they shall have atone on-time of at least 200 milliseconds.

2. Self-recording audiometers shall comply with the following requirements:

(A) The chart upon which the audiogram is traced shall have lines at positions corresponding to all multiples of 10 dB hearing level within the intensity range spanned by the audiometer. The lines shall be equally spaced and shall be separated by at least 1/4 inch. Additional increments are optional. The audiogram pen tracings shall not exceed 2 dB in width.

(B) It shall be possible to set the stylus manually at the 10-dB increment lines for calibration purposes.

(C) The slewing rate for the audiometer attenuator shall not be more than 6 dB/sec except that an initial slewing rate greater than 6 dB/sec is permitted at the

Appendix C: Occupational Noise Exposure

beginning of each new test frequency, but only until the second subject response.

(D) The audiometer shall remain at each required test frequency for 30 seconds (+ or - 3 seconds). The audiogram shall be clearly marked at each change of frequency and the actual frequency change of the audiometer shall not deviate from the frequency boundaries marked on the audiogram by more than + or - 3 seconds.

(E) It must be possible at each test frequency to place a horizontal line segment parallel to the time axis on the audiogram, such that the audiometric tracing crosses the line segment at least six times at that test frequency. At each test frequency the threshold shall be the average of the midpoints of the tracing excursions.

1910.95 Appendix D—Audiometric Test Rooms

Part Number:	1910
Part Title:	Occupational Safety and Health Standards
Subpart:	G
Subpart Title:	Occupational Health and Environment Control
Standard Number:	1910.95 App D
Title:	Audiometric test rooms

This Appendix is Mandatory

Rooms used for audiometric testing shall not have background sound pressure levels exceeding those in Table D-1 when measured by equipment conforming at least to the Type 2 requirements of American National S tandard Specification for Sound Level Meters, S1.4-1971 (R1976), and to the Class II requirements of American National Standard Specification for Octave, Half-Octave, and Third-Octave Band Filter Sets, S1.11-1971 (R1976).

Table D-1. Maximum Allowable Octave-Band Sound Pressure Levels for Audiometric Test Rooms

Octave-band center frequency (Hz)	500	1000	2000	4000	8000
Sound pressure level (dB)	40	40	47	57	62

1910.95 Appendix E—Acoustic Calibration of Audiometers

Part Number:	1910
Part Title:	Occupational Safety and Health Standards
Subpart:	G
Subpart Title:	Occupational Health and Environment Control
Standard Number:	1910.95 App E
Title:	Acoustic calibration of audiometers

This Appendix is Mandatory

Audiometer calibration shall be checked acoustically, at least annually, according to the procedures described in this appendix. The equipment necessary to perform these measurements is a sound level meter, octave-band filter set, and a National Bureau of Standards 9A coupler. In making these measurements, the accuracy of the calibrating equipment shall be sufficient to determine that the audiometer is within the tolerance permitted by American Standard Specification for Audiometers, S3.6-1969.

(1) "Sound Pressure Output Check"

A. Place the earphone coupler over the microphone of the sound level meter and place the earphone on the couple.
B. Set the audiometer's hearing threshold level (HTL) dial to 70 dB.
C. Measure the sound pressure level of the tones at each test frequency from 500 Hz through 6000 Hz for each earphone.
D. At each frequency the readout on the sound level meter should correspond to the levels in Table E-1 or Table E-2, as appropriate, for the type of earphone, in the column entitled "sound level meter reading."

(2) "Linearity Check"

A. With the earphone in place, set the frequency to 1000 Hz and the HTL dial on the audiometer to 70 dB.
B. Measure the sound levels in the coupler at each 10-dB decrement from 70 dB to 10 dB, noting the sound level meter reading at each setting.
C. For each 10-dB decrement on the audiometer the sound level meter should indicate a corresponding 10 dB decrease.
D. This measurement may be made electrically with a voltmeter connected to the earphone terminals.

(3) "Tolerances"

When any of the measured sound levels deviate from the levels in Table E-1 or Table E-2 by + or - 3 dB at any test frequency between 500 and 3000 Hz, 4 dB at 4000

Appendix C: Occupational Noise Exposure

Hz, or 5 dB at 6000 Hz, an exhaustive calibration is advised. An exhaustive calibration is required if the deviations are greater than 15 dB or greater at any test frequency.

Table E-1. Reference Threshold Levels for Telephonics—TDH-39 Earphones

Frequency, Hz	Reference threshold level TDH-39 earphones, dB	Sound level meter reading, dB
500	11.5	81.5
1000	7	77
2000	9	79
3000	10	80
4000	9.5	79.5
6000	15.5	85.5

Table E-2. Reference Threshold Levels for Telephonics—TDH-49 Earphones

Frequency, Hz	Reference threshold level TDH-39 earphones, dB	Sound level meter reading, dB
500	13.5	83.5
1000	7.5	77.5
2000	11	81.0
3000	9.5	79.5
4000	10.5	80.5
6000	13.5	83.5

1910.95 Appendix F—Calculations and Application of Age Corrections to Audiograms

Part Number:	1910
Part Title:	Occupational Safety and Health Standards
Subpart:	G
Subpart Title:	Occupational Health and Environment Control
Standard Number:	910.95 App F
Title:	Calculations and application of age corrections to audiograms

This Appendix Is Non-Mandatory

In determining whether a standard threshold shift has occurred, allowance may be made for the contribution of aging to the change in hearing level by adjusting the most recent audiogram. If the employer chooses to adjust the audiogram, the employer shall follow the procedure described below. This procedure and the age correction tables were developed by the National Institute for Occupational Safety and Health in the criteria document entitled "Criteria for a Recommended Standard . . . Occupational Exposure to Noise," ((HSM)-11001).

For each audiometric test frequency:

(i) Determine from Tables F-1 or F-2 the age correction values for the employee by:
   (A) Finding the age at which the most recent audiogram was taken and recording the corresponding values of age corrections at 1000 Hz through 6000 Hz;
(B) Finding the age at which the baseline audiogram was taken and recording the corresponding values of age corrections at 1000 Hz through 6000 Hz.

(ii) Subtract the values found in step (i)(B) from the value found in step (i)(A).

(iii) The differences calculated in step (ii) represented that portion of the change in hearing that may be due to aging.

EXAMPLE:

Employee is a 32-year-old male. The audiometric history for his right ear is shown in decibels below.

The audiogram at age 27 is considered the baseline since it shows the best hearing threshold levels. Asterisks have been used to identify the baseline and most recent audiogram. A threshold shift of 20 dB exists at 4000 Hz between the audiograms taken at ages 27 and 32.

(The threshold shift is computed by subtracting the hearing threshold at age 27,

Appendix C: Occupational Noise Exposure

	Audiometric test frequency (Hz)				
Employee's Age	1000	2000	3000	4000	6000
26	10	5	5	10	5
*27	0	0	0	5	5

which was 5, from the hearing threshold at age 32, which is 25). Are test audiogram has confirmed this shift. The contribution of aging to this change in hearing may be estimated in the following manner:

Go to Table F-1 and find the age correction values (in dB) for 4000 Hz at age 27 and age 32.

The difference represents the amount of hearing loss that may be attributed to aging in the time period between the baseline audiogram and the most recent audiogram. In this example, the difference at 4000 Hz is 3 dB. This value is subtracted from the hearing level at 4000 Hz, which in the most recent audiogram is 25, yielding 22 after adjustment. Then the hearing threshold in the baseline audiogram at 4000 Hz (5) is subtracted from the adjusted annual audiogram hearing threshold at 4000 Hz (22). Thus the age-corrected threshold shift would be 17 dB (as opposed to a threshold shift of 20 dB without age correction).

	Frequency (Hz)				
Employee's Age	1000	2000	3000	4000	6000
*Age 32	6	5	7	10	14
Age 27	5	4	6	7	11
Difference	1	1	1	3	3

Table F-1. Age Correction Values in Decibels for Males

| Years | \multicolumn{5}{c|}{Audiometric Test Frequency (Hz)} |
	1000	2000	3000	4000	6000
20 or younger	5	3	4	5	8
21	5	3	4	5	8
22	5	3	4	5	8
23	5	3	4	6	9
24	5	3	5	6	9
25	5	3	5	7	10
26	5	4	5	7	10
27	5	4	6	7	11
28	6	4	6	8	11
29	6	4	6	8	12
30	6	4	6	9	12
31	6	4	7	9	13
32	6	5	7	10	14
33	6	5	7	10	14
34	6	5	8	11	15
35	7	5	8	11	15
36	7	5	9	12	16
37	7	6	9	12	17
38	7	6	9	13	17
39	7	6	10	14	18
40	7	6	10	14	19
41	7	6	10	14	20
42	8	7	11	16	20
43	8	7	12	16	21
44	8	7	12	17	22
45	8	7	13	18	23
46	8	8	13	19	24
47	8	8	14	19	24
48	9	8	14	20	25
49	9	9	15	21	26
50	9	9	16	22	27
51	9	9	16	23	28
52	9	10	17	24	29
53	9	10	18	25	30
54	10	10	18	26	31
55	10	11	19	27	32
56	10	11	20	28	34
57	10	11	21	29	35
58	10	12	22	31	36
59	11	12	22	32	37
60 or older	11	13	23	33	38

Appendix C: Occupational Noise Exposure

Table F-2. Age Correction Values in Decibels for Females

	Audiometric Test Frequency Hz)				
Years	1000	2000	3000	4000	6000
20 or younger	7	4	3	3	6
21	7	4	4	3	6
22	7	4	4	4	6
23	7	5	4	4	7
24	7	5	4	4	7
25	8	5	4	4	7
26	8	5	5	4	8
27	8	5	5	5	8
28	8	5	5	5	8
29	8	5	5	5	9
30	8	6	5	5	9
31	8	6	6	5	9
32	9	6	6	6	10
33	9	6	6	6	10
34	9	6	6	6	10
35	9	6	7	7	11
36	9	7	7	7	11
37	9	7	7	7	12
38	10	7	7	7	12
39	10	7	8	8	12
40	10	7	8	8	13
41	10	8	8	8	13
42	10	8	9	9	13
43	11	8	9	9	14
44	11	8	9	9	14
45	11	8	10	10	15
46	11	9	10	10	15
47	11	9	10	11	16
48	12	9	11	11	16
49	12	9	11	11	16
50	12	10	11	12	17
51	12	10	12	12	17
52	12	10	12	13	18
53	13	10	13	13	18
54	13	11	13	14	19

(Continued on page 159)

(Continued from page 158)

Table F-2—Continued

Years	\multicolumn{5}{c}{Audiometric Test Frequency(Hz)}
	1000	2000	3000	4000	6000
55	13	11	14	14	19
56	13	11	14	15	20
57	13	11	15	15	20
58	14	12	15	16	21
59	14	12	16	16	21
60 or older	14	12	16	17	22

1910.95 Appendix G—Monitoring Noise Levels

Non-mandatory informational appendix

Part Number:	1910
Part Title:	Occupational Safety and Health Standards
Subpart:	G
Subpart Title:	Occupational Health and Environment Control
Standard Number:	1910.95 App G
Title:	Monitoring noise levels non-mandatory informational appendix

This appendix provides information to help employers comply with the noise monitoring obligations that are part of the hearing conservation amendment.

WHAT IS THE PURPOSE OF NOISE MONITORING?

This revised amendment requires that employees be placed in a hearing conservation program if they are exposed to average noise levels of 85 dB or greater during an 8 hour workday. In order to determine if exposures are at or above this level, it may be necessary to measure or monitor the actual noise levels in the workplace and to estimate the noise exposure or "dose" received by employees during the workday.

WHEN IS IT NECESSARY TO IMPLEMENT A NOISE MONITORING PROGRAM?

It is not necessary for every employer to measure workplace noise. Noise monitoring or measuring must be conducted only when exposures are at or above 85 dB. Factors which suggest that noise exposures in the workplace may be at this level include employee complaints about the loudness of noise, indications that employees are losing their hearing, or noisy conditions which make normal conversation difficult. The employer should also consider any information available regarding noise emitted from specific machines. In addition, actual workplace noise measurements can suggest whether or not a monitoring program should be initiated.

HOW IS NOISE MEASURED?

Basically, there are two different instruments to measure noise exposures: the sound level meter and the dosimeter. A sound level meter is a device that measures the intensity of sound at a given moment. Since sound level meters provide a measure of sound intensity at only one point in time, it is generally necessary to take a number of measurements at different times during the day to estimate noise exposure over a workday. If noise levels fluctuate, the amount of time noise remains at each of the various measured levels must be determined.

To estimate employee noise exposures with a sound level meter it is also generally necessary to take several measurements at different locations within the workplace. After appropriate sound level meter readings are obtained, people sometimes draw "maps" of the sound levels within different areas of the workplace. By using a sound level "map" and information on employee locations throughout the day, estimates of individual exposure levels can be developed. This measurement method is generally referred to as "area" noise monitoring.

A dosimeter is like a sound level meter except that it stores sound level measurements and integrates these measurements over time, providing an average noise exposure reading for a given period of time, such as an 8-hour workday. With a dosimeter, a microphone is attached to the employee's clothing and the exposure measurement is simply read at the end of the desired time period. A reader may be used to read-out the dosimeter's measurements. Since the dosimeter is worn by the employee, it measures noise levels in those locations in which the employee travels. A sound level meter can also be positioned within the immediate vicinity of the exposed worker to obtain an individual exposure estimate. Such procedures are generally referred to as "personal" noise monitoring.

Area monitoring can be used to estimate noise exposure when the noise levels are relatively constant and employees are not mobile. In workplaces where employees move about in different areas or where the noise intensity tends to fluctuate over time, noise exposure is generally more accurately estimated by the personal monitoring approach.

In situations where personal monitoring is appropriate, proper positioning of the microphone is necessary to obtain accurate measurements. With a dosimeter, the microphone is generally located on the shoulder and remains in that position for the entire workday. With a sound level meter, the microphone is stationed near the employee's head, and the instrument is usually held by an individual who follows the employee as he or she moves about.

Manufacturer's instructions, contained in dosimeter and sound level meter operating manuals, should be followed for calibration and maintenance. To ensure accurate results, it is considered good professional practice to calibrate instruments before and after each use.

Appendix C: Occupational Noise Exposure

HOW OFTEN IS IT NECESSARY TO MONITOR NOISE LEVELS?

The amendment requires that when there are significant changes in machinery or production processes that may result in increased noise levels, remonitoring must be conducted to determine whether additional employees need to be included in the hearing conservation program. Many companies choose to remonitor periodically (once every year or two) to ensure that all exposed employees are included in their hearing conservation programs.

WHERE CAN EQUIPMENT AND TECHNICAL ADVICE BE OBTAINED?

Noise monitoring equipment may be either purchased or rented. Sound level meters cost about $500 to $1,000, while dosimeters range in price from about $750 to $1,500. Smaller companies may find it more economical to rent equipment rather than to purchase it. Names of equipment suppliers may be found in the telephone book (Yellow Pages) under headings such as: "Safety Equipment," "Industrial Hygiene," or "Engineers-Acoustical." In addition to providing information on obtaining noise monitoring equipment, many companies and individuals included under such listings can provide professional advice on how to conduct a valid noise monitoring program. Some audiological testing firms and industrial hygiene firms also provide noise monitoring services. Universities with audiology, industrial hygiene, or acoustical engineering departments may also provide information or may be able to help employers meet their obligations under this amendment.

Free, on-site assistance may be obtained from OSHA-supported state and private consultation organizations. These safety and health consultative entities generally give priority to the needs of small businesses.

1910.95 Appendix H—Availability of Referenced Documents

Part Number:	1910
Part Title:	Occupational Safety and Health Standards
Subpart:	G
Subpart Title:	Occupational Health and Environment Control
Standard Number:	1910.95 App H
Title:	Availability of referenced documents

Paragraphs (c) through (o) of 29 CFR 1910.95 and the accompanying appendices contain provisions which incorporate publications by reference. Generally, the publications provide criteria for instruments to be used in monitoring and audiometric testing. These criteria are intended to be mandatory when so indicated in the applicable paragraphs of 1910.95 and appendices.

It should be noted that OSHA does not require that employers purchase a copy of the referenced publications. Employers, however, may desire to obtain a copy of the referenced publications for their own information.

The designation of the paragraph of the standard in which the referenced publications appear, the titles of the publications, and the availability of the publications are as follows:

Paragraph designation	Referenced publication	Available from—
Appendix B…..	"List of Personal Hearing Protectors and Attenuation Data, " HEW Pub. No. 76-120, 1975. NTISW-PB267461.	National Technical Information Service, Port Royal Road, Springfield, VA 22161
Appendix D…..	"Specification for Sound Level Meters," S1.4-1971 (R1976)	American National Standards Institute, Inc., 1430 Broadway, New York, NY 10018
1910.95(k) (2), Appendix E	"Specifications for Audiometers," S3.6-1969.	American National Standards Institute, Inc., 1430 Broadway, New York, NY 10018

Appendix C: Occupational Noise Exposure

The referenced publications (or a microfiche of the publications) are available for review at many universities and public libraries throughout the country. These publications may also be examined at the OSHA Technical Data Center, Room N2439, United States Department of Labor, 200 Constitution Avenue, NW., Washington, DC 20210, (202) 219-7500 or at any OSHA Regional Office (see telephone directories under United States Government - Labor Department).

[61 FR 9227, March 7, 1996]

1910.95 Appendix I—Definitions

Part	Number:1910
Part Title:	Occupational Safety and Health Standards
Subpart:	G
Subpart Title:	Occupational Health and Environment Control
Standard Number:	1910.95 App I
Title:	Definitions

These definitions apply to the following terms as used in paragraphs (c) through (n) of 29 CFR 1910.95.

- Action level - An 8-hour time-weighted average of 85 decibels measured on the A-scale, slow response, or equivalently, a dose of fifty percent.

- Audiogram - A chart, graph, or table resulting from an audiometric test showing an individual's hearing threshold levels as a function of frequency.

- Audiologist - A professional, specializing in the study and rehabilitation of hearing, who is certified by the American Speech-Language-Hearing Association or licensed by a state board of examiners.

- Baseline audiogram - The audiogram against which future audiograms are compared.

- Criterion sound level - A sound level of 90 decibels.

- Decibel (dB) - Unit of measurement of sound level.Hertz (Hz) - Unit of measurement of frequency, numerically equal to cycles per second.

- Medical pathology - A disorder or disease. For purposes of this regulation, a condition or disease affecting the ear, which should be treated by a physician specialist.

- Noise dose - The ratio, expressed as a percentage, of (1) the time

Otolaryngologist - A physician specializing in diagnosis and treatment of disorders of the ear, nose and throat.

Representative exposure - Measurements of an employee's noise dose or 8-hour time-weighted average sound level that the employers deem to be representative of the exposures of other employees in the workplace.

Sound level - Ten times the common logarithm of the ratio of the square of the measured A-weighted sound pressure to the square of the standard reference pressure of 20 micropascals. Unit: decibels (dB). For use with this regulation, SLOW time response, in accordance with ANSI S1.4-1971 (R1976), is required.

Sound level meter - An instrument for the measurement of sound level.

Time-weighted average sound level - That sound level, which if constant over an 8-hour exposure, would result in the same noise dose as is measured.

[39 FR 23502, June 27, 1974, as amended at 46 FR 4161, Jan. 16, 1981; 46 FR 62845, Dec. 29, 1981; 48 FR 9776, Mar. 8, 1983; 48 FR 29687, June 28, 1983; 54 FR 24333, June 7, 1989; 61 FR 5507, Feb. 13, 1996; 61 FR 9227, March 7, 1996]

Appendix C: Occupational Noise Exposure

Appendix D

Part 1904—OSHA Recordkeeping Related to Hearing Loss Cases

Appendix B - Federal Registers

Occupational Injury and Illness Recording and Reporting Requirements -- Final Rule - 67:44037-44048

Publication Date:	07/01/2002
Publication Type:	Final Rules
Fed Register #:	67:44037-44048
Standard Number:	1904; 1910.95; 1910.1020; 1952
Title:	Occupational Injury and Illness Recording and Reporting Requirements -- Final Rule

PART 1904 -- [AMENDED]

1. The authority citation for part 1904 continues to read as follows:

Authority: 29 U.S.C. 657, 658, 660, 666, 673, Secretary of Labor's Order No. 3-2000 (65 FR 50017), and 5 U.S.C. 533.

2. Revise § 1904.10 to read as follows:

§ 1904.10 Recording criteria for cases involving occupational hearing loss.

(a) Basic requirement. If an employee's hearing test (audiogram) reveals that the employee has experienced a work-related Standard Threshold Shift (STS) in hearing in one or both ears, and the employee's total hearing level is 25 decibels (dB) or more above audiometric zero (averaged at 2000, 3000, and 4000 Hz) in the same ear(s) as the STS, you must record the case on the OSHA 300 Log.

(b) Implementation.

(1) What is a Standard Threshold Shift? A Standard Threshold Shift, or STS, is defined in the occupational noise exposure standard at 29 CFR 1910.95(g)(10)(i) as a change in hearing threshold, relative to the baseline audiogram for that employee, of an average of 10 decibels (dB) or more at 2000, 3000, and 4000 hertz (Hz) in one or both ears.

Appendix D: Part 1904, OSHA Recordkeeping Related to Hearing Loss Cases

(2) How do I evaluate the current audiogram to determine whether an employee has an STS and a 25-dB hearing level?

(i) STS. If the employee has never previously experienced a recordable hearing loss, you must compare the employee's current audiogram with that employee's baseline audiogram. If the employee has previously experienced a recordable hearing loss, you must compare the employee's current audiogram with the employee's revised baseline audiogram (the audiogram reflecting the employee's previous recordable hearing loss case).

(ii) 25-dB loss. Audiometric test results reflect the employee's overall hearing ability in comparison to audiometric zero. Therefore, using the employee's current audiogram, you must use the average hearing level at 2000, 3000, and 4000 Hz to determine whether or not the employee's total hearing level is 25 dB or more.

(3) May I adjust the current audiogram to reflect the effects of aging on hearing?

Yes. When you are determining whether an STS has occurred, you may age adjust the employee's current audiogram results by using Tables F-1 or F-2, as appropriate, in Appendix F of 29 CFR 1910.95. You may not use an age adjustment when determining whether the employee's total hearing level is 25 dB or more above audiometric zero.

(4) Do I have to record the hearing loss if I am going to retest the employee's hearing?

No, if you retest the employee's hearing within 30 days of the first test, and the retest does not confirm the recordable STS, you are not required to record the hearing loss case on the OSHA 300 Log. If the retest confirms the recordable STS, you must record the hearing loss illness within seven (7) calendar days of the retest. If subsequent audiometric testing performed under the testing requirements of the § 1910.95 noise standard indicates that an STS is not persistent, you may erase or line-out the recorded entry.

(5) Are there any special rules for determining whether a hearing loss case is work-related?

No. You must use the rules in § 1904.5 to determine if the hearing loss is work-related. If an event or exposure in the work environment either caused or contributed to the hearing loss, or significantly aggravated a pre-existing hearing loss, you must consider the case to be work related.

(6) If a physician or other licensed health care professional determines the hearing loss is not work-related, do I still need to record the case?

If a physician or other licensed health care professional determines that the hearing loss is not work-related or has not been significantly aggravated by occupational noise exposure, you are not required to consider the case work-related or to record the case on the OSHA 300 Log.

(7) How do I complete the 300 Log for a hearing loss case?

When you enter a recordable hearing loss case on the OSHA 300 Log, you must check the 300 Log column for hearing loss.

Note to 1904.10(b)(7): The applicability of paragraph (b)(7) is delayed until further notice.

[FR Doc. 02-16392 Filed 6-28-02; 8:45 am]

BILLING CODE 4510-26-P

Appendix D: Part 1904, OSHA Recordkeeping Related to Hearing Loss Cases

OSHA Trade News Release

U.S. Department of Labor
Office of Public Affairs

TRADE NEWS RELEASE
Monday, December 16, 2002
Contact: Frank Meilinger
Phone: (202) 693-1999

OSHA RECORKEEPING FORM TO INCLUDE HEARING LOSS IN 2004
MSD Decisions Delayed

WASHINGTON -- Beginning Jan. 1, 2004, employers will be required to check a hearing loss column to record work-related cases meeting the new recording criteria established by the Occupational Safety and Health Administration. The new criteria go into effect in 2003.

"The new recordkeeping standard requires employers to record work-related hearing loss cases when an employee's hearing test shows a marked decrease in overall hearing," said OSHA Administrator John Henshaw. "Data from the new column will improve the nation's statistical information on occupational hearing loss, improve the agency's ability to determine where the injuries occur, and help prioritize hearing loss prevention efforts."

Under the new criteria, employers will record 10-decibel shifts from the employee's baseline hearing test when they also result in an overall hearing level of 25 decibels.

OSHA is also postponing for one year three provisions related to musculoskeletal disorders (MSDs); the rule's definition of musculoskeletal disorders (MSDs), consideration of MSDs as privacy concern cases, and requirements to check a MSD columns on the OSHA Log.

The delay does not effect an employer's obligation to record workplace injuries and illnesses or keep workplaces free from hazards. However, employers will not be required to use an MSD definition to categorize cases on the OSHA Log for calendar year 2003. Instead, they must check the column for "injury" or "all other illness" depending on the circumstances of the case.

OSHA also clarified three matters relating to recording occupational hearing loss in conjunction with the final rule: audiometric tests for workers in the shipbuilding industry; computation of a standard threshold shift for determining recordable hearing loss, and how OSHA will treat an expected increase in the number of recorded cases resulting from new recordkeeping definitions requirements.

Information on OSHA's decision to delay the effective date of the recordkeeping provisions and clarification on recording occupational hearing loss is scheduled for publication in the December 19 *Federal Register*.

OSHA will announce its decision on the need for an MSD column in a future Federal Register document.

The Occupational Safety and Health Administration is dedicated to saving lives, preventing injuries and illnesses, and protecting America's workers. Safety and health add value to business, the workplace and life. For more information, visit http://www.osha.gov/index.html.

###

Appendix E

Directory of OSHA Regional Offices

Region 1

Connecticut, Massachusetts, Maine, New Hampshire, Rhode Island, Vermont
   Regional Office
   JFK Federal Building, Room E340
   Boston, Massachusetts 02203
   (617) 565-9860
   (617) 565-9827 FAX

Region 2

New Jersey, New York, Puerto Rico, Virgin Islands
   Regional Office
   201 Varick Street, Room 670
   New York, New York 10014
   (212) 337-2378
   (212) 337-2371 FAX

Region 3

District of Columbia, Delaware, Maryland, Pennsylvania, Virginia, West Virginia
   U.S. Department of Labor/OSHA
   The Curtis Center-Suite 740 West
   170 S. Independence Mall West
   Philadelphia, PA 19106-3309
   (215) 861-4900
   (215) 861-4904 FAX

Region 4

Alabama, Florida, Georgia, Kentucky, Mississippi, North Carolina, South Carolina, Tennessee

Appendix E: Directory of OSHA Regional Offices

Regional Office
61 Forsyth Street, SW
Atlanta, Georgia 30303
(404) 562-2300
(4040 562-2295 FAX

Region 5
Illinois, Indiana, Michigan, Minnesota, Ohio, Wisconsin
Regional Office
230 South Dearborn Street, Room 3244
Chicago, Illinois 60604
(312) 353-2220
(312) 353-7774 FAX

Region 6
Arkansas, Louisiana, New Mexico, Oklahoma, Texas
Regional Office
525 Griffin Street, Room 602
Dallas, Texas 75202
(214) 767-4731
(214) 767-4137 FAX

Region 7
Iowa, Kansas, Missouri, Nebraska
Regional Office
City Center Square
1100 Main Street, Suite 800
Kansas City, Missouri 64105
(816) 426-5861
(816) 426-2750 FAX

Region 8
Colorado, Montana, North Dakota, South Dakota, Utah, Wyoming
Regional Office

1999 Broadway, Suite 1690
P.O. Box 46550
Denver, Colorado 80201-6550
(303) 844-1600
(303) 844-1616 FAX

Region 9
Arizona, California, Guam, Hawaii, Nevada
Region IX Federal Contact Numbers
71 Stevenson Street, Room 420
San Francisco, California 94105
(415) 975-4310 (Main Public – 8:00 AM – 4:30 PM Pacific)
(800) 475-4019 (For Technical Assistance)
(800) 475-4020 (For Complaints – Accidents/Fatalities)
(800) 475-4022 (For Publication Requests)
(415) 975-4319 FAX

Region 10
Alaska, Idaho, Oregon, Washington
Regional Office
1111 Third Avenue, Suite 715
Seattle, Washington 98101-3212
(206) 553-5930
(206) 553-6499 FAX

Appendix E: Directory of OSHA Regional Offices

Appendix F

Directory of States with Approved Occupational Safety and Health Plans

Last Updated: 15 July 2002

Alaska Department of Labor and Workforce Development
P.O. Box 21149
1111 W. 8th Street, Room 306
Juneau, Alaska 99802-1149
Ed Flanagan, Commissioner (907) 465-2700 Fax: (907) 465-2784
Richard Mastriano, Program Director (907) 269-4904 Fax: (907) 269-4915

Industrial Commission of Arizona
800 W. Washington
Phoenix, Arizona 85007-2922
Larry Etchechury, Director, ICA (602) 542-4411 Fax: (602) 542-1614
Darin Perkins, Program Director (602) 542-5795 Fax: (602) 542-1614

California Department of Industrial Relations
455 Golden Gate Avenue - 10th Floor
San Francisco, California 94102
Steve Smith, Director (415) 703-5050 Fax:(415) 703-5114
Dr. John Howard, Chief (415) 703-5100 Fax: (415) 703-5114
Ray Yee, Manager, Cal/OSHA Program Office (415) 703-5177 Fax: (415) 703-5114

Connecticut Department of Labor
200 Folly Brook Boulevard
Wethersfield, Connecticut 06109
Shaun Cashman, Commissioner (860) 566-5123 Fax: (860) 566-1520
Conn-OSHA

Appendix F: Directory of States with Approved OS&H Plans

38 Wolcott Hill Road
Wethersfield, Connecticut 06109
Donald Heckler, Director (860) 566-4550 Fax: (860) 566-6916

Hawaii Department of Labor and Industrial Relations
830 Punchbowl Street
Honolulu, Hawaii 96813
Leonard Agor, Director (808) 586-8844 Fax: (808) 586-9099
Jennifer Shishido, Administrator (808) 586-9116 Fax: (808) 586-9104

Indiana Department of Labor
State Office Building
402 West Washington Street, Room W195
Indianapolis, Indiana 46204-2751
John Griffin, Commissioner (317) 232-2378 Fax: (317) 233-3790
John Jones, Deputy Commissioner (317) 232-3325 Fax: (317) 233-3790

Iowa Division of Labor
1000 E. Grand Avenue
Des Moines, Iowa 50319-0209
Byron K. Orton, Commissioner (515) 281-6432 Fax: (515) 281-4698
Mary L. Bryant, Administrator (515) 242-5870 Fax: (515) 281-7995

Kentucky Labor Cabinet
1047 U.S. Highway 127 South, Suite 4
Frankfort, Kentucky 40601
Joe Norsworthy, Secretary (502) 564-3070 Fax: (502) 564-5387
William Ralston, Federal\State Coordinator (502) 564-3070 ext.240 Fax: (502) 564-1682

Maryland Division of Labor and Industry
Department of Labor, Licensing and Regulation
1100 North Eutaw Street, Room 613
Baltimore, Maryland 21201-2206
Kenneth P. Reichard, Commissioner (410) 767-2241 Fax: (410) 767-2986

Keith Goddard, Deputy Commissioner (410) 767-2992 Fax: (410) 767-2986
Cheryl Kammerman, Assistant Commissioner, MOSH (410) 767-2215 Fax: (410) 333-7747

Michigan Department of Consumer and Industry Services
Kathleen M. Wilbur, Director
Bureau of Safety and Regulation
P.O. Box 30643
Lansing, MI 48909-8143
Douglas R. Earle, Director (517) 322-1814 Fax: (517) 322-1775
Doug Kalinowski, Deputy Director for Enforcement
(517) 322-1817 Fax: (517) 322-1775

Minnesota Department of Labor and Industry
443 Lafayette Road
St. Paul, Minnesota 55155
Shirley Chase, Commissioner (651) 284-5010 Fax: (651) 282-5405
Rosyln Wade, Assistant Commissioner (651) 284-5018 Fax: (651) 282-5293
Patricia Todd, Administrative Director, OSHA Management Team
(651) 284-5372 Fax: (651) 297-2527

Nevada Division of Industrial Relations
400 West King Street, Suite 400
Carson City, Nevada 89703
Roger Bremmer, Administrator (775) 687-3032 Fax: (775) 687-6305
Occupational Safety and Health Enforcement Section (OSHES)
1301 N. Green Valley Parkway
Henderson, Nevada 89014
Tom Czehowski, Chief Administrative Officer (702) 486-9168 Fax:(702) 990-0358
[Las Vegas (702) 687-5240]

New Jersey Department of Labor
John Fitch Plaza - Labor Building
Market and Warren Streets
P.O. Box 110
Trenton, New Jersey 08625-0110

Appendix F: Directory of States with Approved OS&H Plans

Albert G. Kroll, Commssioner (609) 292-2975 Fax: (609) 633-9271
Leonard Katz, Assistant Commissioner (609) 292-2313 Fax: (609) 695-1314
Howard Black, Assistant Director (609) 292-2425 Fax: (609) 292-3749

New Mexico Environment Department
1190 St. Francis Drive
P.O. Box 26110
Santa Fe, New Mexico 87502
Peter Maggiore, Secretary (505) 827-2850 Fax: (505) 827-2836
Sam A. Rogers, Chief (505) 827-4230 Fax: (505) 827-4422

New York Department of Labor
New York Public Employee Safety and Health Program
State Office Campus Building 12, Room 158
Albany, New York 12240
Linda Angello, Commissioner (518) 457-2746 Fax: (518) 457-6908
Richard Cucolo, Director, Division of Safety and Health
(518) 457-3518 Fax: (518) 457-1519
Maureen Cox, Program Manager (518) 457-1263 Fax: (518) 457-5545

North Carolina Department of Labor
4 West Edenton Street
Raleigh, North Carolina 27601-1092
Cherie Berry, Commissioner (919) 807-2900 Fax: (919) 807-2855
John Johnson, Deputy Commissioner, OSH Director (919) 807-2861 Fax: (919) 807-2855
Kevin Beauregard, OSH Assistant Director (919) 807-2863 Fax:(919) 807-2856

Oregon Occupational Safety and Health Division
Department of Consumer & Business Services
350 Winter Street, NE, Room 430
Salem, Oregon 97310-0220
Peter DeLuca, Administrator (503) 378-3272 Fax: (503) 947-7461
David Sparks, Deputy Administrator for Policy (503) 378-3272 Fax: (503) 947-7461
Michele Patterson, Deputy Administrator for Operations (503) 378-3272
   Fax: (503) 947-7461

Puerto Rico Department of Labor and Human Resources
Prudencio Rivera Martínez Building
505 Muñoz Rivera Avenue
Hato Rey, Puerto Rico 00918
Víctor Rivera Hernández, Secretary
(787) 754-2119 Fax: (787) 753-9550
Brenda Sepúlveda, Assistant Secretary for Occupational Safety and Health
(787) 756-1100, 1106 / 754-2171 Fax: (787) 767-6051
José Droz, Deputy Director for Occupational Safety and Health
(787) 756-1100, 1106 / 754-2188 Fax: (787) 767-6051

South Carolina Department of Labor, Licensing, and Regulation
Koger Office Park, Kingstree Building
110 Centerview Drive
PO Box 11329
Columbia, South Carolina 29211
Rita McKinney, Director (803) 896-4300 Fax: (803) 896-4393
Michelle Childs, Deputy Director (803) 734-4248

Tennessee Department of Labor
710 James Robertson Parkway
Nashville, Tennessee 37243-0659
Michael E. Magill, Commissioner (615) 741-2582 Fax: (615) 741-5078
John Winkler, Acting Program Director (615) 741-2793 Fax: (615) 741-3325

Utah Labor Commission
160 East 300 South, 3rd Floor
PO Box 146650
Salt Lake City, Utah 84114-6650
R. Lee Ellertson, Commissioner (801) 530-6901 Fax: (801) 530-7906
Larry Patrick, Administrator (801) 530-6898 Fax: (801) 530-6390

Vermont Department of Labor and Industry
National Life Building - Drawer 20
Montpelier, Vermont 05620-3401

Appendix F: Directory of States with Approved OS&H Plans

Tasha Wallis, Commissioner (802) 828-2288 Fax: (802) 828-2748
Robert McLeod, Project Manager (802) 828-2765 Fax: (802) 828-2195

Virgin Islands Department of Labor
3021 Golden Rock
Christiansted, St. Croix, Virgin Islands 00820-4660
Cecil R. Benjamin, Commissioner (340) 773-1994 Fax: (340) 773-1858
John Sheen, Assistant Commissioner (340) 772-1315 Fax: (340) 772-4323
Francine Lang, Program Director (340) 772-1315 Fax: (340) 772-4323

Virginia Department of Labor and Industry
Powers-Taylor Building
13 South 13th Street
Richmond, Virginia 23219
C. Raymond Davenport, Commissioner (804) 786-2377 Fax: (804) 371-6524
Jay Withrow, Director, Office of Legal Support (804) 786-9873 Fax: (804) 786-8418
Glenn Cox, Director, Safety Compliance, VOSHA (804) 786-2391 Fax: (804) 371-6524

Washington Department of Labor andIndustries
General Administration Building
PO Box 44001
Olympia, Washington 98504-4001
Gary Moore, Director (360) 902-4200 Fax: (360) 902-4202
Michael Silverstein, Assistant Director [PO Box 44600]
(360) 902-5495 Fax: (360) 902-5529
Steve Cant, Program Manager, Federal-State Operations [PO Box 44600]
(360) 902-5430 Fax: (360) 902-5529

Wyoming Department of Employment
Workers' Safety and Compensation Division
Herschler Building, 2nd Floor East
122 West 25th Street
Cheyenne, Wyoming 82002
Stephan R. Foster, Safety Administrator (307) 777-7786
Fax: (307) 777-3646

Appendix G

Directory of OSHA Consultation Offices

Updated 26 June 2002

ALABAMA
Safe State Program
University of Alabama
432 Martha Parham West
P.O. Box 870388
Tuscaloosa, Alabama 35487
(205) 348-3033
(205) 348-3049 FAX
E-mail: bweems@ccs.ua.edu
Website: http://bama.ua.edu/~deip/safe_state_osha.html

ALASKA
Consultation Section, ADOL/AKOSH
3301 Eagle Street
P.O. Box 107022
Anchorage, Alaska 99510
(907) 269-4957
(907) 269-4950 FAX
E-mail: cliff_hustead@labor.state.ak.us
Website: http://www.labor.state.ak.us/lss/oshhome.html

ARIZONA
Consultation & Training
Industrial Commission of Arizona
Division of Occupational Safety & Health
800 West Washington
Phoenix, Arizona 85007-9070
(602) 542-1695
(602) 542-1614 FAX
E-mail: pat.ryan@osha.gov

ARKANSAS
OSHA Consultation
Arkansas Department of Labor
10421 West Markham
Little Rock, Arkansas 72205
(501) 682-4522
(501) 682-4532 FAX
E-mail: clark.thomas@osha.gov

Appendix G: Directory of OSHA Consultation Offices

Website: http://www.state.ar.us/labor/serv01.html

CALIFORNIA
CAL/OSHA Consultation Service
Department of Industrial Relations
2424 Arden Way, Suite 485
Sacramento, California 95825
(916) 263-5765
(916) 263-5768FAX
E-mail: InfoCons@hq.dir.ca.gov
Website: http://www.dir.ca.gov/DOSH/consultation.htm

COLORADO
Colorado State University
Occupational Safety & Health Section
115 Environmental Health Building
Fort Collins, Colorado 80523
(970) 491-6151
(970) 491-7778 FAX
E-mail: del.sandfort@lamar.colostate.edu
Website: http://www.bernardino.colostate.edu/enhealth/7c1.htm

CONNECTICUT
Connecticut Department of Labor
Division of Occupational Safety & Health
38 Wolcott Hill Road
Wethersfield, Connecticut 06109
(860) 566-4550
(860) 566-6916 FAX
E-mail: donald.heckler@osha.gov
Website: http://www.ctdol.state.ct.us/osha/osha.html

DELAWARE
Delaware Department of Labor
Division of Industrial Affairs
Occupational Safety & Health
4425 Market Street
Wilmington, Delaware 19802
(302) 761-8219
(302) 761-6601 FAX
E-mail: ttrznadel@state.de.us
Website: http://www.state.de.us/labor/aboutdol/industrialaffairs.htm

DISTRICT OF COLUMBIA
Program available only for employers within the District of Columbia
DC Department of Employment Services
Office of Occupational Safety & Health
77 P Street, N.E., 2nd Floor
Washington, D.C. 20002
(202)671-1800
(202) 671-3018 FAX
E-mail: jcates@osha.gov

FLORIDA
University of South Florida
Safety Florida Consultation Program
Department of Environmental & Occupational Health
College of Public Health
4003 East Fowler Avenue
Tampa, Florida 33617
(813) 974-9962
(813) 974-9973FAX
E-mail: cvespi@hsc.usf.edu
Website: http://www.safetyflorida.usf.edu

GEORGIA
Georgia Institute of Technology
21(d) Onsite Consultation Program
151 6th Street, NW
O'Keefe Building, Room 025
Atlanta, Georgia 30332-0837
(404) 894-8276
(404) 894-8275 FAX
E-mail: daniel.ortiz@gtri.gatech.edu
Website: http://www.oshainfo.gatech.edu

GUAM
OSHA Onsite Consultation
Dept. of Labor, Government of Guam
107 F Street
Tiyam, Guam 96931
(671) 475-0136
(671) 477-9503FAX
E-mail: terrybadley@osha.gov
Website: http://mail.admin.gov.gu/webdol/oshacompl.html

HAWAII
Consultation & Training Branch
Department of Labor & Industrial Relations
830 Punchbowl Street
Honolulu, Hawaii 96813
(808) 586-9100
(808) 586-9104 FAX
E-mail: ellen.kondo@osha.gov
Website: http://www.state.hi.us/dlir/hiosh/consult.html

IDAHO
Boise State University Occupational Safety & Health Program
1910 University Drive
Boise, Idaho 83725-1825
(208) 426-3283
(208) 426-4411 FAX
E-mail: lhill@boisestate.edu
Website: http://www2.boisestate.edu/ehs/Consultation.html

Appendix G: Directory of OSHA Consultation Offices

ILLINOIS
Illinois Onsite Consultation
Industrial Service Division
Department of Commerce & Community Affairs
State of Illinois Center, Suite 3-400
100 West Randolph Street
Chicago, Illinois 60601
(312) 814-2337
(312) 814-7238 FAX
E-mail: sfryzel@commerce.state.il.us
Website: http://www.commerce.state.il.us/Services/SmallBusiness/OSHA/OSHAhome.html

INDIANA
Bureau of Safety, Education & Training
Division of Labor, Room W195
402 West Washington
Indianapolis, Indiana 46204-2287
(317) 232-2688
(317) 232-3790 FAX
E-mail: cmack@col.state.in.us
Website: http://www.state.in.us/labor

IOWA
Iowa Workforce Development & Labor Services
Bureau of Consulation & Education
1000 East Grand
DesMoines, Iowa 50319
(515) 281-7629
(515) 281-5522 FAX
E-mail: steve.slater@osha.gov
Website: http://www.state.ia.us/iwd/labor/index.html

KANSAS
Kansas 7(c)(1) Consultation Program
Kansas Dept. of Human Resources
512 South West 6th Street
Topeka, Kansas 66603-3150
(785) 296-4386
(785) 296-1775 FAX
E-mail: rudy.leutzinger@osha.gov

KENTUCKY
Kentucky Labor Cabinet
Division of Education & Training
Kentucky OSH Program
1047 U.S. Highway 127, South
Frankfort, Kentucky 40601
(502) 564-3070
(502) 564-4769FAX
E-mail: arussell@mail.lab.state.ky.us
Website: http://www.kylabor.net/kyosh/oshcons.html

LOUISIANA
7(c)(1) Consultation Program
Louisiana Department of Labor
1001 N. 23rd Street, Room 230
P.O. Box 94094
Baton Rouge, Louisiana 70804-9094
(225) 342-9601
(225) 342-5158 FAX
E-mail: cmills@ldol.state.la.us

MAINE
Maine Bureau of Labor Standards
Workplace Safety & Health Division
State House Station #45
Augusta, Maine 04333-0045
(207) 624-6463
(207) 624-6449 FAX
E-mail: david.e.wacker@state.me.us
Website: http://janus.state.me.us/labor/consult.html

MARYLAND
MOSH Consultation Services
312 Marshall Avenue, Room 600
Laurel, Maryland 20707
(410) 880-6131
(301) 880-6369FAX
E-mail: andrew.alcarese@osha.gov
Website: http://www.dllr.state.md.us/labor/mosh.htm

MASSACHUSETTS
Division of Occupaional Safety & Health
Dept. of Workforce Development
1001 Watertown Street
West Newton, Massachusetts 02165
(617) 727-3982
(617) 727-4581 FAX
E-mail: joe.lamalva@state.ma.us
Website: http://www.state.ma.us/dos/Consult/Consult.html

MICHIGAN
Department of Consumer & Industry Services
7150 Harris Drive
Lansing, Michigan 48909
(517) 322-1809
(517) 322-1374 FAX
E-mail: ayalew.kanno@cis.state.mi.us
Website: http://www.cis.state.mi.us/bsr/divisions/set/set_con.html

MINNESOTA
Department of Labor & Industry
Consultation Division
443 LaFayette Road

Appendix G: Directory of OSHA Consultation Offices

Saint Paul, Minnesota 55155
(651) 284-5060
(651) 297-1953(FAX)
E-mail: james.collins@state.mn.us
Website: http://www.doli.state.mn.us/mnosha.htm

MISSISIPPI
Mississippi State University
Center for Safety & Health
106 Crosspark Drive
Suite C
Pearl, Mississippi 39208
601-939-2047
601-939-6742 FAX
E-mail: kelly.tucker@osha.gov
Website: http://www.msstate.edu/dept/csh

MISSOURI
Onsite Consultation Program
Division of Labor Standards
Dept. of Labor & Industrial Relations
3315 West Truman Boulevard
Post Office Box 449
Jefferson City, Missouri 65109
(573) 751-3403
(573) 751-3721 FAX
E-mail: laborstandards@dolir.state.mo.us
Website: http://www.dolir.state.mo.us/ls/onsite/index.htm

MONTANA
Department of Labor & Industry
Bureau of Safety
PO Box 1728
Helena, Montana 59624-1728
(406) 444-6418
(406) 444-9396FAX
E-mail: smahalik@state.mt.us
Website: http://erd.dli.state.mt.us/Safety/SBhome.html

NEBRASKA
Nebraska Workforce Development
Office of Safety & Labor Standards
State Office Building, Lower Level
301 Centennial Mall, South
Lincoln, Nebraska 68509-5024
(402) 471-4717
(402) 471-5039 FAX
E-mail: ediedrichs@dol.state.ne.us
Website: http://www.dol.state.ne.us/safety/7c1.html

NEVADA
Safety Consultation & Training Section

Division of Industrial Relations
Department of Business & Industry
1301 Green Valley Parkway
Henderson, Nevada 89074
(702) 486-9140
(702) 990-0362 FAX
E-mail: jan.rosenberg@osha.gov
Website: http://4safenv.state.nv.us

NEW HAMPSHIRE
New Hampshire Dept of Health & Human Services
6 Hazen Drive
Concord, New Hampshire 03301-6527
(603) 271-2024
(603) 271-2667 FAX
E-mail: stephen.beyer@osha.gov
Website:http://www.dhhs.state.nh.us/CommPublicHealth/oshcs.nsf/vMain

NEW JERSEY
New Jersey Department of Labor
Division of Public Safety & Occupational Safety & Health
225 E. State Street
8th Floor West
P.O. Box 953
Trenton, New Jersey 08625-0953
(609) 292-3923
(609) 292-4409 FAX
E-mail: carol.farley@osha.gov
Website: http://www.state.nj.us/labor/consult.htm

NEW MEXICO
New Mexico Environment Department
Occupational Health & Safety Bureau
525 Camino de Los Marquez, Suite 3
PO Box 26110
Santa Fe, New Mexico 87502
(505) 827-4230
(505) 827-4422 FAX
E-mail: Kevin_Koch@nmenv.state.nm.us
Website: http://www.nmenv.state.nm.us/env_prot.htm

NEW YORK
Division of Safety & Health
State Office Campus
Building 12, Room 130
Albany, New York 12240
(518) 457-2238
(518) 457-3454 FAX
E-mail: james.rush@osha.gov
Website: http://www.labor.state.ny.us/html/employer/p469.htm

NORTH CAROLINA
Bureau of Consultative Services

Appendix G: Directory of OSHA Consultation Offices

NC Department of Labor--OSHA Division
4 West Edenton Street
Raleigh, North Carolina 27601-1092
(919) 807-2899
(919) 807-2902 FAX
E-mail: wjoyner@mail.dol.state.nc.us
Website: http://www.dol.state.nc.us/osha/consult/consult.html

NORTH DAKOTA
North Dakota Department of Health/Consolidated Labs
Environmental Health Sectioni
1200 Missouri Avenue, Room 304
Bismarck, North Dakota 58504
(701) 328-5188
(701) 328-5200 FAX
E-mail: agilliss@state.nd.us
Website: http://www.ehs.health.state.nd.us/ndhd/environ/ee/oshc/index.html

OHIO
On-Site Consultation Program
Bureau of Occupational Safety & Health
LAWS Division / Ohio Dept. of Commerce
50 W. Broad Street, Suite 2900
Columbus, Ohio 43215
1-800-282-1425 or 614-644-2631
614-644-3133 FAX
E-mail: wes.hohl@perrp.com.state.oh.us
Website: http://198.234.41.214/w3/webpo2.nsf

OKLAHOMA
Oklahoma Department of Labor
OSHA Division
4001 North Lincoln Boulevard
Oklahoma City, Oklahoma 73105-5212
(405) 528-1500
(405) 528-5751 FAX
E-mail: diana.jones1@osha.gov
Website: http://www.okdol.state.ok.us/osha

OREGON
Oregon OSHA
Department of Consumer & Business Services
350 Winter Street, N.E., Room 430
Salem, Oregon 97310
(503) 378-3272
(503) 378-5729 FAX
E-mail: michelle.cattanach@state.or.us
Website: http://www.orosha.org

PENNSYLVANIA
Indiana University Pennsylvania
Room 210 Walsh Hall

302 East Walk
Indiana, Pennsylvania 15705-1087
(724) 357-2561
(724) 357-2385 FAX
E-mail: john.engler@osha.gov
Website: http://www.iup.edu/sa/osha/index.htm

PUERTO RICO
Occupational Safety & Health Office
Department of Labor & Human Resources, 21st Floor
505 Munoz Rivera Avenue
Hato Rey, Puerto Rico 00918
(787) 754-2171
(787) 767-6051 FAX
E-mail: mvelez@osha.gov

RHODE ISLAND
OSH Consultation Program
Division of Occupational Health & Radiation Control
Rhode Island Department of Health
3 Capital Hill
Providence, Rhode Island 02908
(401) 222-2438
(401) 222-2456 FAX
E-mail: safesite@doh.state.ri.us
Website: http://www.state.ri.us/dohrad.html

SOUTH CAROLINA
South Carolina Department of Labor, Licensing & Regulation
3600 Forest Drive
P.O. Box 11329
Columbia, South Carolina 29204
(803) 734-9614
(803) 734-9741 FAX
E-mail: bob.peck@osha.gov
Website: http://www.llr.state.sc.us/oshavol.html

SOUTH DAKOTA
South Dakota State University
Engineering Extension
West Hall, Box 510
907 Harvey Dunn Street
Brookings, South Dakota 57007-0597
(605) 688-4101
(605) 688-6290 FAX
E-mail: james_manning@sdstate.edu

TENNESSEE
OSHA Consultation Services Division
Tennessee Department of Labor
3rd floor, Andrew Johnson Tower
710 James Robertson Parkway
Nashville, Tennessee 37243-0659

Appendix G: Directory of OSHA Consultation Offices

(615) 741-7155
(615) 532-2997 FAX
E-mail: jcothron@mail.state.tn.us
Website: http://www.state.tn.us/labor/toshcons.htm

TEXAS
Workers' Health & Safety Division
Texas Workers' Compensation Commission
Southfield Building
4000 South I H 35
Austin, Texas 78704
(512) 804-4640
(512) 804-4601FAX
OSHCON Request Line: 800-687-7080
E-mail: jimmy.harper@twcc.state.tx.us
Website: http://twcc.state.tx.us/services/oshcon.htm

UTAH
State of Utah Labor Commission
Workplace Safety & Health
Consultation Services
160 East 300 South
Salt Lake City, Utah 84114-6650
(801) 530-6901
(801) 530-6992 FAX
E-mail: icmain.nanderso@state.ut.us
Website: Click Here

VERMONT
Vermont Department of Labor & Industry
Division of Occupational Safety & Health
National Life Building, Drawer 20
Montpelier, Vermont 05602-3401
(802) 828-2765
(802) 828-2195 FAX
E-mail: robert.mcleod@labind.state.vt.us
Website: http://www.state.vt.us/labind/vosha.html

VIRGINIA
Virginia Department of Labor & Industry
Occupational Safety & Health
Training & Consultation
13 South 13th Street
Richmond, Virginia 23219
(804) 786-6359
(804) 786-8418 FAX
E-mail: wer@doli.state.va.us
Website: http://www.dli.state.va.us/programs/consultation.html

VIRGIN ISLANDS
Division of Occupational Safety & Health
Virgin Islands Department of Labor
3021 Golden Rock
Christiansted

St. Croix, Virgin Island 00840
(340) 772-1315
(340) 772-4323 FAX
Website: http://www.gov.vi/vild

WASHINGTON
Washington Dept of Labor & Industries
WISHA Services Division
P.O. Box 44649
Olympia, Washington 98504
(360) 902-5443
(360) 902-5459 FAX
E-mail: jame235@lni.wa.gov
Website: http://www.wa.gov/lni/wisha/wisha.html

WEST VIRGINIA
West Virginia Department of Labor
Capitol Complex Building #6
1800 East Washington Street, Room B-749
Charleston, West Virginia 25305
(304) 558-7890
(304) 558-3797FAX
E-mail: jburgess@labor.state.wv.us
Website: http://www.state.wv.us/labor/sections.html

WISCONSIN (Health)
Wisconsin Department of Health & Family Services
Division of Public Health
1 West Wilson Street
Room B157
Madison, WI 53701-2659
(608) 266-9383
(608) 266-1550FAX
E-mail: terry.moen@osha.gov
Website: http://www.dhfs.state.wi.us/dph_boh/OSHA_Cons/index.html

WISCONSIN (Safety)
Wisconsin Department of Commerce
Bureau of Marketing, Advocacy & Technology Development
Bureau of Manufacturing & Assessment
N14 W23833 Stone Ridge Drive Suite B100
Waukesha, Wisconsin 53188-1125
(262) 523-3044 -800-947-0553
(262) 523-3046 FAX
E-mail: jim.lutz@osha.gov
Website: http://www.commerce.state.wi.us/MT/MT-FAX-0928.htm

WYOMING
Wyoming Department of Employment
Workers' Safety & Compensation Division
Herschler Building, 2 East
122 West 25th Street
Cheyenne, Wyoming 82002